Sustainable Development

Sustainable Development

Understanding the green debates

Mark Mawhinney

Sustainable Cities Research Institute
University of Northumbria at Newcastle

Blackwell
Science

© 2002 by Blackwell Science Ltd,
a Blackwell Publishing Company
Editorial Offices:
Osney Mead, Oxford OX2 0EL, UK
  Tel: +44 (0)1865 206206
Blackwell Science Inc., 350 Main Street,
Malden, MA 02148-5018, USA
  Tel: +1 781 388 8250
Iowa State Press, a Blackwell Publishing
Company, 2121 State Avenue, Ames,
Iowa 50014-8300, USA
  Tel: +1 515 292 0140
Blackwell Publishing Asia Pty Ltd, 550 Swanston
Street, Carlton South, Melbourne, Victoria 3053,
Australia
  Tel: +61 (0)3 9347 0300
Blackwell Wissenschafts Verlag,
Kurfürstendamm 57, 10707 Berlin, Germany
  Tel: +49 (0)30 32 79 060

First published 2002 by Blackwell Science Ltd

Library of Congress
Cataloging-in-Publication Data
is available

ISBN 0-632-06459-5

A catalogue record for this title is available
from the British Library

Set in 10/13pt Palatino
by DP Photosetting, Aylesbury, Bucks
Printed and bound in Great Britain by
MPG Books Ltd, Bodmin, Cornwall

For further information on
Blackwell Science, visit our website:
www.blackwell-science.com

Contents

1 An Overview

Sustainable development as a concept promises many things to many people. Aspects of government policy, business strategy and even lifestyle decisions have been shaped around the concept. However, there is still an ambiguity surrounding the subject and the meaning of the words themselves.

The phrase 'sustainable development' has been continually redefined to cover ever-growing parts of life on the planet, and some of the definitions are shown overleaf. From early 'green' definitions which concentrated on environmental concerns, the definitions quickly moved on to cover ever-wider issues, raising the possibility of conflicting principles, compromise and doubts on whether anything can ever be agreed. In short, it has become a complex interdisciplinary subject providing an interesting case study of the constraints and pitfalls in modern living. *Is there, therefore, a common definition of sustainable development which applies to all cases?*

The purpose of this book is straightforward: to simply describe some of the various strands of thought that each purport to define sustainable development as a concept, theory or set of principles or processes. After describing such thoughts, the focus is switched to the gaps and the difficulties in understanding that arise in debates on the subject. The final chapters then provide a review of what is needed to break down the existing barriers to a common understanding and attempt to redefine the current concepts in terms that may provide a means of progressing the debate on the subject.

Sustainable development is a complex subject which is difficult to encapsulate in one book, and some of the description and analysis inevitably introduces limitations:

- In the early chapters some of the schools of thought are grouped in a controversial fashion, necessary to allow coverage of a full spectrum of views.
- Many of the arguments provided are generally rooted in a *developed world* view of sustainable development. Views from

less wealthy developing nations can be very different, and many texts on the subject would argue the need to consider both at the same time.

■ Much of the text is framed around four questions, deemed as critical to the understanding of the wider concept. They are introduced here in this first chapter and assessed again in later sections, before a final review in Chapter 8. The first question provides the framework for introducing the subject in this first chapter, while the other three gradually enter into the discussion across the chapter:

(1) Is sustainable development a concept that defines a starting point, or does it define the process necessary, or should it be the defining end-goal?

(2) Does the concept of sustainable development provide a coherent theory?

(3) Is sustainable development a workable concept in practice?

(4) Is it 'balanced' or does 'balance' form a part of the solution?

1.2   Sustainable development: the starting point?

There are many accepted or acceptable definitions of sustainable development and therein lies the first set of problems; is it possible to have one defining explanation, or does it depend on your political viewpoint? More importantly, do the definitions provide a starting point, process or end-goal?

The following provides a flavour of the various descriptions, explanations or definitions available. It is difficult to provide an accurate, fair sample simply because there are so many definitions, and each new presentation on the subject seems to bring a new or refined definition. However, the intention in presenting the list is to push the reader into (1) assessing a range of views without removing any at this stage as so often happens, and (2) assessing whether the objective in setting each of the definitions is to establish a starting point, process or end-goal.

The first in the list is the most noted definition, which came from the Brundtland report, a landmark document in the sustainable development debate which is further explored here and again in Chapter 3. However, there are many definitions, often associated with one particular group of players in the debate, and the details of each of the major groupings of schools of thought will be studied in the next few chapters.

What is sustainable development?

Brundtland (1987)
'Sustainable development is development that meets the needs of the present without compromising the ability of future generations to meet their own needs.'

National Strategies for Sustainable Development (2000)
Sustainable development is 'economic and social development that meets the needs of the current generation without undermining the ability of future generations to meet their own needs.'

World Wildlife Fund (IUCN et al. 1991)
'Sustainable development means improving the quality of life while living within the carrying capacity of supporting systems.'

ICLEI (International Council for Local Environmental Initiatives) 1994
Sustainable development 'delivers basic environmental, social and economic services to all residents of a community without threatening the viability of the natural, built and social systems upon which the delivery of these services depends'.

LGMB (Local Government Management Board, UK) 1993
Sustainable development is 'reducing current levels of consumption of energy and resources and production of waste in order not to damage the natural systems which future generations will rely on to provide them with resources, absorb their waste and provide safe and healthy living conditions'.

UK Department of Environment, Transport and Regions (1999a)
- *'Social progress that recognises the needs of everyone*
- *Effective protection of the environment*
- *Prudent use of natural resources*
- *Maintenance of high and stable level of economic growth and employment.'*

US Department of Energy (2001)
'Sustainable development is a strategy by which communities seek economic development approaches that also benefit the local environment and quality of life. It has become an important guide to many communities that have discovered that traditional approaches to plan-

ning and development are creating, rather than solving, societal and environmental problems.'

Schoonbrodt (1995)
Sustainability should include:
- *'all rounded development, economic, social, cultural and political*
- *equal rights for all with the best quality of life to each and every person*
- *reject social, economic and political exclusion*
- *control pollution and minimize waste*
- *pleasure of city life, dismissing the "back to nature" dream'.*

Novartis Foundation for Sustainable Development (2001)
Sustainable development involves 'Programmes in the developing countries that directly contribute to an improvement in the quality of life of the poorest people'.

Wackernagel and Rees (1996)
Sustainable development is 'The need for humanity to live equitably within the means of nature'.

Robert et al. (1997)
'A compass for sustainable development:
- *Does an action cause a decrease on use of metals, fuels and minerals?*
- *Does an action increase dependence on unnatural substances?*
- *Does an activity encroach on productive parts of nature?*
- *Does an activity result in use of unnecessarily large amounts of resources?'*

Pearce et al. (1990)
Sustainable development means that 'conditions necessary for equal access to the resource base be met for each generation'.

World Bank (Pezzey 1989)
'Sustainable development will be non-declining per capita utility – because of its self-evident appeal as a criterion for intergenerational equity.'

What is sustainable business practice?

World Business Council for Sustainable Development (2001)
Sustainable business practice requires 'Business leadership as a catalyst for change toward sustainable development, and to promote the role of eco-efficiency, innovation, and corporate social responsibility toward sustainable development'.

What is a sustainable city?

Girardet (1999)
'A sustainable city is organised so as to enable its citizens to meet their own needs and to enhance their well-being without damaging the natural world or endangering the living conditions of other people, now or in the future.'

Thus, sustainable development appears to be an over-used, misunderstood phrase. It is often presented as a mission statement (a starting point perhaps?), at a time when there is general recognition of mission statement fatigue, and there are different interpretations from a variety of business, policy-makers, the health sector and academics.

If it was the case that there was no one explicit definition then there would be no common agenda, just a range of views using the same language. However, some further study of the sample definitions suggests that there may be common issues which underpin many of them. Three of the definitions are therefore tested in more detail in order to look for key components of the debate, the range of factors that are accepted, the evidence underpinning them and to define the challenges that arise from accepting the components and the definitions. This can lead to prioritisation of the key factors and identification of what is needed in a good universal definition, if such a concept is possible.

Definition 1

Development that meets the needs of the present without compromising the ability of future generations to meet their own needs (Brundtland 1987). Intergenerational legacy and the need to limit development to only that which is a necessity are the key elements of this definition, i.e., avoid passing problems onto the future and avoid wasting resources. At this level it is a clear, highly principled message with an emphasis on conservation and ensuring that future generations can enjoy the same breadth of choices as current generations.

It is a simple message but its interpretation can become difficult. It spans everything from an individual becoming accountable for living within their means through to the global community living within the sustainable resource of the world, and consequently covers a huge breadth of activity and scale of activity. There are further difficulties with the definition of the limits and constraints of the resource available, since a message of conservation can often imply that any change, no matter how necessary, is bad.

The Brundtland definition finds most support in major global organisations. However, organisations such as the World Bank, the European Union (EU) and the Organisation for Economic Co-operation and Development (OECD) tend to flesh out the statement with a series of caveats and defining objectives. The OECD approach, for example, is set out in the second definition in the list (National Strategies for Sustainable Development 2000), clearly stating economic and social development as key factors, with no mention of the environment.

By contrast the World Bank's definition looks to a balance of five environmental, five social and five economic needs or priorities underpinning the definition (World Bank 2001), suggesting a different set of priorities.

Definition 2

Social progress that recognises the needs of everyone, effective protection of the environment, prudent use of natural resources and maintenance of high and stable levels of economic growth and employment. A commonly used definition in the United Kingdom (UK) was developed by the national government's Department of Environment, Transport and Regions (1999a). The department had a broad remit of activity and, as such, its choice of definition followed a strategy of maximum coverage, but with little focus to highlight the trade-offs necessary (Cantle 1999).

The attraction in this definition is its inclusion of key-words, pointing quite clearly at three critical main elements: social, environmental and economic issues. At this level it thus appears to be directional in nature, rather than principled (although the principles are well defined in other parts of the same reference).

The difficulty with this definition is that words like 'high', 'prudent', 'effective' or even 'recognises' are all ambiguous. Again, there are key problems in establishing appropriate scale and particular limitations, important factors that often explain the current frustrating desire for redefinition of sustainability for every new project.

The UK government approach

The Department of Environment, Transport and Regions (DETR) approach is an interesting case study, showing how the case for sustainable development has been made. Throughout the 1990s the DETR was involved in studying, monitoring and development of guidance on sustainable development. In 1999 this culminated in two documents designed to define sustainable development for national and local government policy. The first document, 'A better quality of life' (DETR 1999a), presented the arguments for a balanced approach to cover economic, environmental and social issues. The study looked at principles through to indicators and implications. The key overall objective was to 'ensure a better quality of life for everyone, now and for generations to come'.

A second document (DETR 1999b) presented the detail and evidence in support of the arguments. This report advocated the use of 15 headline indicators and 150 general indicators in order to present full coverage of the three sets of issues in a broad manner and to provide a benchmark for influencing future national and local government targets. An impressive set of data was assembled as a baseline for the position in 1999, showing progress being made in some areas and deterioration in others.

The headline indicators were split into four areas of coverage as follows:

(1) *Maintaining high and stable levels of economic growth* – GDP/ GDP per head, total and social investment as a percentage of GDP, proportion of people of working age in work.
(2) *Social progress which recognises the needs of everyone* – success in tackling poverty and social inclusion, average qualifications at age 19, expected years of healthy life, homes judged unfit for habitation, levels of crime.
(3) *Effective protection of the environment* – emissions of greenhouse gases, days when air pollution is moderate or higher, road traffic, rivers of good or fair quality, populations of birds, new homes built on previously developed land.
(4) *Prudent use of natural resources* – waste arising and management.

The choice of the headline indicators and the 150 general indicators was influenced by three main factors:

(1) International obligations and influential work elsewhere, most notably by the OECD producing a core set of 40–50

indicators in 1991, the United Nations producing a set of 134 indicators in 1996 and the European Union's development of 60 environmental indicators in 1999.

(2) The need to produce SMART (suitable, measurable, achievable, realistic and time-constrained) measurable targets at both national and local government levels to match these indicators.

(3) The need to consider the concept of economic capital (saving and investment in the financial sense), social capital (skills, knowledge, health and social networks that can be utilised to generate work and well-being) and environmental capital (protecting the diversity and abundance of nature), their interactions and the hierarchies that are most important.

The key factors are provided as checklists rather than presenting them as an interactive model. These checklists lend themselves to an indicator approach in evaluating sustainable development, and this is further discussed in Chapter 6.

The DETR noted, however, that there are models which have been used to draw together similar work elsewhere which include the United Nation's Human Development Index (see Chapter 4), on-going work on some form of Green National Product, where GDP is amended to include environmental damage arising from the wealth production (see Chapter 2), and more complicated 'pressure, state, response' models (see Chapter 5).

Definition 3

A sustainable city is organised so as to enable its citizens to meet their own needs and to enhance their well-being without damaging the natural world or endangering the living conditions of other people, now or in the future. The third definition examined is one provided by Girardet (1999) aimed specifically at cities. Cities are increasingly being perceived as a key factor in defining and delivering sustainable development. They are viewed as being a primary cause of the concentration of bad practice which leads to unsustainable development. City-dwellers see few of the worst consequences of their consumption of resource and their concentration in one location can often exacerbate problems if development is not sustainable.

Girardet's definition emphasises people and their long-term future. It also notes the need to avoid damaging the environment. He sees the concept as encompassing local needs and local decision-making and, at the same time, being aware of the effects of the local on the planet. He is optimistic in that he advocates improvement rather than removal of cities.

Thus, Brundtland appears to be a statement of principle, the UK government definition identifies the elements that may make up the subject and Girardet suggests a mix of the two in his definition. All three definitions stress the importance of humanity in the equation.

To further test the commonality and differences in definitions, Table 1.1 takes 7 of the earlier 17 listed and highlights the critical elements of the definitions, the apparent objectives and the difficulties of application in each case. Analysis of the table reveals a number of common themes; the importance of a long-term view, an agreement on some sort of balance between economic, environmental and social needs and the inclusion of quality of life or poverty as an issue. These are therefore the priorities needed in a universal definition and become the common denominators for all work on sustainable development. However, they raise a number of potential difficulties: the risks of predicting the future, an acceptance of complex inter-disciplinary work, a tension between development for human needs and its effect on environment, and the suggestion that it encompasses many issues which may hinder any plans for any future change or development.

Do the definitions of sustainable development lend themselves to becoming a starting point for theory and for practice? For it to be a good starting point there would need to be one set of clear universal principles. The evidence produced in the text to date suggests that there is not sufficient clarity in definition to create one clear, simple starting point. However, it has some promise in the apparent broad agreement on principles and this will be explored further in later chapters.

Scale, complexity and compromise have been highlighted as the issues that cause greatest difficulty in seeking a universal definition. The next section therefore looks at the potential to define sustainble development as a theory that defines a process.

Table 1.1 The main components of some definitions.

Definition	Message	General objectives	Difficulties
Brundtland	Intergenerational legacy Constraint development	Development on needs only with minimal damage basis	How do you measure needs of the future? Does not address scale?
National Strategies for Sustainable Development	Socio-economic development Intergenerational legacy	Similar to Brundtland but narrower base	How do you measure needs of the future? Does not address scale?
UK Department of Environment, Transport and Regions	Social progress, economic growth, environmental and resource protection	Balance of interests	Compromise and conflict – who decides priorities?
Girardet	Citizen need and well-being Environmental protection	Equity and avoid damage to others	Who organises the operating system?
Wackernagel and Rees	Equitable living Environmental protection	Acknowledge the limits of resources in equitable manner	Who decides and who organises? – evidence base?
Robert *et al.*	Limits to natural resources	Acknowledge the limits of resources	What are the socio-economic effects of this?
Pearce *et al.*	Equal access to resources across generations	Acknowledge the limits of resources in equitable intergenerational manner	Who decides and organises? – evidence base?

1.3   Sustainable development as a process

It has been noted that the older, greener arguments, which saw sustainable development as concerned solely with environmental issues, have now been replaced by rounder, fuller versions with consideration of social, economic and environmental aspects of life. From a political point of view this has been useful since it has allowed a wider audience to embrace sustainable development, beyond the earlier devotees who may have become viewed as radical and disruptive (Murray 2001). However, it begs the question of how to develop a unifying, coherent theme to cover all of the ever-widening scope of issues. Acceptance of this widely accepted notion of sustainable development still requires much thought on a suitable end-goal and the methodology or process of reaching that end-goal (Fig. 1.1).

Fig. 1.1   The path to sustainable development.

For anybody embarking on research or study in the area it quickly becomes obvious that the starting point is different for the various professions representing the environmental, the economic and the social angle. They speak in different languages and have different priorities. This in turn leads them to take different approaches and to head towards differing end-points. Krupp (1996) states, for example, that development is generally accepted by economists to mean demographic and economic growth, i.e., a successful city is viewed as one that is growing with a healthy, long-living population who become ever richer. Environmentalists would, however, wish to acknowledge the constraints of nature and impose restraint on growth.

There are therefore many contradictions in studying sustainable development across subject disciplines as well as across different levels of scales. There are a number of options for the study of such complex subjects:

(1) *Break the subject down into smaller subjects and study it element by element in a scientific manner.* This appears to have been the starting point for the current debate as ecologists split out their view from the more mainstream approach, which was often dominated by economists. However, as stated previously, sustainable development became redefined as the sum of the parts when the splitting out process itself became identified as the root cause of many problems (this is studied in more detail in Chapter 7).

(2) *Develop a set of assumptions which creates a composite political picture and from this develop an ideal set of behaviour as an objective.* Much of the on-going research, which is studied in Chapters 2–5, is developed in this manner, but this again presents problems since different starting points present differing process and end-goals.

(3) *Define the starting point, define the end (balance) point and map out the process to get there (Fig. 1.1).* This is an undoubted area of weakness in the development of a theory of sustainable development to date, seldom considered in texts on the subject. It is, however, the mode by which many practical or implementation studies proceed.

The debate on theory, definition and principles has often been conducted through the first two types of approach. These methods, however, are problematic, since they can often introduce a bias and they seldom exhibit 'joined-up' thinking between professions. The lack of a unifying, universally accepted theory has therefore caused a vacuum which forces practitioners to redefine the starting point, the end-point and the process for every new project, and this is further discussed in Chapter 5.

Option three, the preference of practitioners rather than theoreticians, does not appear to have been fully tested as a basis for a better or coherent theory, and it is this approach that forms the basis of the approach within this text. A methodology based on this approach would need to address the important factor that, as an end-goal, all players need to see some form of payback as their ultimate goal, rather than there being one right answer (from approach one) or one group winning out over others (from approach two). Mapping the range of paybacks across a range of interests and creating a route to that end-goal should raise awareness of the bias that each player is likely to bring to the initial picture.

While it may not be possible to develop a 'universal' process map, examination of option three should help identify real constraints to the development of a proper theoretical picture. Figure 1.2 provides a process map of the steps involved. The most widely used option for practitioners to carve out a path to progress or success has been through the development of a set of locally derived indicators, and their subsequent application based on a locally defined set of acceptable targets. This implies a jump to defining a set of end-goals before starting-point or process have been developed, and some further discussion at the start of Chapter 6 reveals that this is the case.

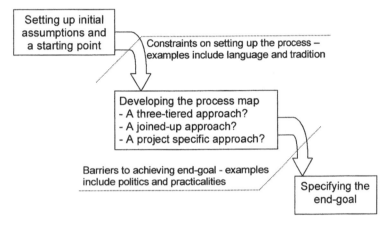

Fig. 1.2   The steps to developing a process map.

The apparent aim for many of these enthusiasts is to achieve some sort of balance in the consideration between social, economic and environmental issues in development, and the indicators act as proxies for the end-goal. However, this raises many questions. How can such a balance be defined and how would a successful outcome be measured? Is a balance truly desirable in any case? How can the set of indicators produced avoid the bias of their authors?

Unfortunately it is often difficult to see the link in practical studies between the commonly accepted principles of sustainable development and the many sets of locally derived indicators, a subject that will be further studied in Chapter 6. This implies that the process to link starting-point and end-goal is poorly developed. For example, at governmental level there is often a genuine desire to deal with sustainable development, but results can be confusing. Guidance on the subject is seldom uniform, reflecting the difficulties of striking a balance on neutrality, practicality and finding the appropriate scale for judgement.

Figure 1.3 is a graphical illustration of guidance issued by the UK government's Department of Environment, Transport and Regions to a variety of agencies at local or regional level. This appears to indicate a lack of consistency. This leaves a very open definition for the practitioner, and hence the use of proxies and non-standard sets of indicators. While redefining the definitions before each project is useful in the short term it has the unfortunate consequence of simply redefining the problem without progressing the global issue.

Number of indicators

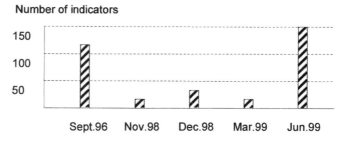

Fig. 1.3 Sustainable development – 'the official view?' (After Mawhinney 1999).

The issue of a traceable, repeatable process that properly represents sustainable development appears, after this cursory look, to be theoretically difficult. Politics, trade-offs, perception, scale and the practical difficulties all cause problems. It is therefore likely that a universal process map is neither achievable nor desirable.

Many of the best applied examples show a positive combined effect in two of the three key dimensions of economic, social and environmental benefit and it is difficult to claim positive effects in all three directions. It would require a well-defined process where all three can be combined rationally. Application and practice will, therefore, continue to be a driving force in the development of the subject. However, without a universal theory there will continue to be no common thread and so the subject of process is reviewed again in Chapter 5.

1.4   Sustainable development: the end-goal?

Thus, current practice appears to rely heavily on indicators as the proxy for the end-goal. Indicator theory, when properly applied, imposes three chief requirements, in order to justify measure and evaluation of plans or performance (Coates *et al.* 1993):

(1)   Information availability
(2)   Acceptability
(3)   'Neutrality'

The availability of information and the acceptability of that information and the process of analysis can be addressed through careful development of the scope and scale of work for each new project. It is clear, however, from analysis in Chapter 6 that even this is seldom properly addressed.

The real and perceived neutrality of the information and analysis is a key question in the correct development of indicators, particularly in a subject such as sustainable development where there are few completely right or wrong solutions. To counter this problem, group decision-making has often been used as a proxy, acceptable but still involving opinion rather than fact. Thus, the obvious conclusion for many projects is that an element of 'political decision' must be involved in the development of the indicators.

A small, research study at our own research institute was conducted in order to categorise the claim to sustainable development of a number of cities which have often been put forward as good examples in the debate. However, the exercise highlighted many difficulties in the universal acceptance of what was best practice, revealing, for example, the importance of perception, political stance and what constitutes the best form of sustainable development.

The city as part of the debate

Many developed-world cities have had a history of poor environmental management which, with hindsight, can be viewed as unsustainable (*Economist* 1999). However, most have corrected the worst excesses at a later date, in a two-stage process shown in Fig. 1.4. This corrective action cannot be guaranteed, and clearly there remain many examples where the second stage has not yet occurred, but it adds difficulties to the analysis and deciding what are the best examples.

Even cities like Seattle or Curitiba still raise awkward questions. Seattle, viewed by many as an idyllic city which encompassed all of the best of North American development combined with proximity to wonderful natural resources, was once thought to be one of North America's most civil cities (*Economist* 2001). However, the lure of the place and the growth of successful businesses in the area have left it with a recent reputation as a city clogged with traffic, the victim of its own success.

Curitiba is often quoted as best-practice urban planning. It is located in the developing nation of Brazil. In 1990 it had a population of 2.2 million people who, between them, had 500,000 cars. However, 75% of the population travelled by bus to work as the city had built itself along five linear dedicated bus routes (Girardet 1999). While Curitiba's emphasis on public transport is a laudable aim, many would query the value of linear development (often seen as the worst form of urban sprawl) in a green-field location.

Thus, value judgements are a part of the system and need to be acknowledged, or, ideally, separated from the scientific. It is also clear that 'perception' has a role and 'change' inevitably occurs, both of which cause change in themselves.

Using a representation as simple as the triangle quickly reveals a potential danger that the centre of gravity of the triangle will be misinterpreted as the ideal, the point of true balance between economic, social and environmental. This is clearly untrue but yet often appears to be the argument behind

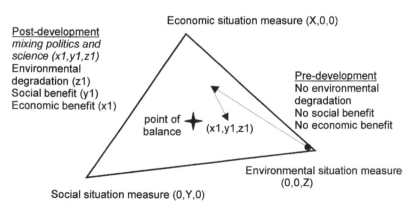

Fig. 1.4 Mapping a development's progress through a triangle.

many of the theories that suggest balance as the best form of sustainable development.

In this respect the representation suggests that sustainable development is no different from current best practice for assessing development (HM Treasury 1997), and it suffers from the same problems of complexity. It is, nevertheless, a useful representation since application to a few case studies rapidly reveals that much development benefits from not being 'perfectly' balanced.

Figure 1.4 indicates how an example project could be mapped in theory across a triangle representing the three directions of social, economic and environmental effects. At a local level any development entails moving from one point representing a set of economic, social and environmental circumstances to another set. The hope of sustainable development is that this movement will be beneficial, i.e., the economic, social and environmental will all improve. In practice, however, this is seldom achievable since there are trade-offs and a perfect balance is indefinable.

Looked at in this manner, there are therefore three variables which vary across the continuum, requiring an optimised solution. The triangle represents a closed set of solutions from which the optimum will be the best combined effect rather than the maximum for any one variable. This is a common type of mathematical problem, an optimisation within a constrained boundary.

Within the triangle the movement involved in any development from pre-development to completion can be mapped, provided suitable proxies can be found for the three variables. The extension of a major out-of-town shopping centre can be used as an example. Pre-development, the environmental effect of the extension would be zero, there would be no economic improvement and no social effect. Thus, the starting point would be the environmental tip of the triangle.

The completed extension is likely to bring economic benefit (measured through, for example, a cost–benefit analysis) and social benefit in the way of new jobs or other quantifiable quality-of-life effects, but it will also deliver environmental degradation through the use of building materials, traffic, energy use, etc. Thus, the point will move towards the interior of the triangle.

There are a number of conclusions from attempting to map the effects in this way. A two-dimensional mapping of this type shows the movement but little else. It assumes, for example, equal weighting of the effects, a linear zero to one across the triangle and agreed measures for each dimension. It cannot address the issues of scale and displacement, once the initial scale has been set. Environmental or social effects which, for example, are displaced elsewhere can be missed if the wrong scale is chosen. The extension of a shopping centre may cause job losses in areas beyond the region of study, a notoriously difficult area to study.

Looking at development in this form of representation it becomes apparent that a perfect balance between environment, social and economic benefit is not definable as suggested earlier. This is examined again in Chapter 5.

The main benefit of practical and applied work has been the development of a set of examples of best practice. Local studies tend to be city, district or project specific and often show examples of restrained resource use, social benefits maximised, economic value maximised or environmental damage limited, although seldom in a full combination.

The main difficulty in establishing end-goals which span across social, economic and environmental issues lies in defining a common currency for the study (Fig. 1.5). The use of indicators suggests that this can be avoided by having a sufficiently rounded set of indicators to fully cover the subject. However, this suggests that the relative improvement of three very different sets of development can be compared without a common currency (Fig. 1.6).

Fig. 1.5   The currency of sustainable development.

In earlier sections the deliberately vague notion of 'development' has been used. However, much depends on the definition of development; is it improvement of mankind, development of the built environment or improvement of the global environment, to

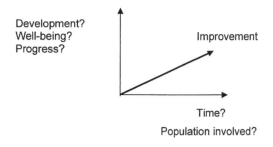

Fig. 1.6   Development or progress?

name but a few of the widely available interpretations? If it is one of these, how is progress or success measured?

In the debate on currency the professionals point in three different directions. The economists point to monetary value, best value or measures of wealth such as Gross Domestic Product (GDP) as the currency of choice. Social scientists suggest that measures of the quality of life (Local Government Association 1999) are critical and environmentalists point to eco-footprints (the misuse or over-use of resource and space) and energy consumption (Wackernagel and Rees 1996).

The city – future and past

It is often argued that cities, as a platform for human habitation, are unsustainable. Much of the argument depends on definitions. At a very simple level, if sustainability is measured in terms of longevity and health (for both the city and its citizens) then most cities have passed the test, and only a small number have withered and died (*Economist* 1999).

If, however, the definition is based around use of resources and destruction of the natural environment then any decision is not so clear-cut. At best, it would appear that humans are reaching a critical point, beyond which usable resources become depleted, bio-diversity suffers and even quality of human life undergoes deterioration. It is unclear how close we are to this point, except in specific cases such as the hole in the ozone layer (DETR 1999b).

It is often argued that city-dwellers, cut off from the negative effects of their decisions, are a major factor that delays action on alleviating such global problems. The relationship between the city and its global effect is not straightforward, and any failure to address the critical issues is likely to become first visible outside city boundaries.

The failure of analysts to convince the inhabitants of cities of this argument highlights a major gap in knowledge. It therefore looks likely that the city is here to stay. The European Commission (1996), for example, promotes the fact that 8 out of 10 Europeans live in towns and cities. Girardet (1999) notes that the global urban population will rise from 50% in the 1990s to 70% in the next century. This trend towards urbanisation arises from the basic human desire to aspire to betterment. The perception that developed-world cities have improved in terms of quality of life across the last 30 years is a strong factor in their attractiveness. It is true that individual cities go through periods of decline and periods of relative rapid improvement, but this cycle has in general led to long-term improvement socially, economically and environmentally, when viewed in terms of human health and quality of life.

The fiction that townsfolk and town authorities actively promote either 'non-betterment' or 'non-sustainability' may serve the purposes of a particular or specialist angle on sustainable development, but it does not serve the common good of cities nor the development of a useful theory on sustainability. Thus, from a human-centred point of view, cities would appear to have proved their worth in the past. It is clear, however, that sustainable development is defined in different ways and at different scales.

Theory and practice therefore need a common language and a common acceptance of the implications of actions within an agreed framework. This must be the main role of sustainable development as an academic subject. However, there are clearly problems of currency, definition, scope and a host of other compromises suggested in this chapter.

Figure 1.7 shows the current three-way split of measures of improvement. The separation of the social, environmental and economic into three factors and their separate measurement provide little clarity. The alternatives, relying on indicators or developing hybrid measures to combine the three (see Chapter 2), represent useful but very imperfect markers.

Voogd (2001), a Dutch planning expert, has provided an interesting hypothesis on the subject. He contends that in the 1970s and 1980s economists influenced the choice of measurement and measured everything in monetary terms but without asking the

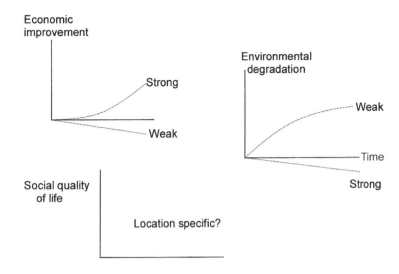

Fig. 1.7   Three components and three measures of progress versus time.

critical question 'who pays?'. In the 1990s environmentalists came to prominence and they demanded the measurement in energy terms. Again, Voogd suggests that they have made the same mistake by not attaching a critical question to the measurement 'what type of energy?'. These astute observations suggest that neither have fully addressed either the issue or the full implications of their own end of the debate.

It may be that the answer is that an envelope of solutions exists for all development alternatives, and this possibility will be examined across the next chapters. However, it is difficult to ignore the initial conclusion that sustainable development is based on part hard fact, part unknown or unmeasured fact and an element of political priority. Can this be quantified? Can it be illustrated through case study? Is it useful to split this out? Can the political or unknown eventually be squeezed out as science improves (Fig. 1.8)?

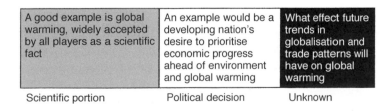

A good example is global warming, widely accepted by all players as a scientific fact	An example would be a developing nation's desire to prioritise economic progress ahead of environment and global warming	What effect future trends in globalisation and trade patterns will have on global warming
Scientific portion	Political decision	Unknown

Fig. 1.8   Splitting out the scientific and the political.

A useful first step for sustainable development experts in the theoretical world, if the above is correct, would be to answer the questions above and find the path between the extremes shown in Fig. 1.7 for a variety of projects, map the common features and improve the arguments. This would make it harder to define a universal answer (as has already been suggested), but what would legitimise any one chosen solution to provide strong, possibly universal, support? The method by which one alternative is taken from the envelope is, again, a political decision.

For example, allowing a target populace to decide what is their priority would be one choice, which would place the advocate in the human-centred end of sustainable development. Others would advocate more scientific or less democratic means. Does the concept of sustainable development lend itself to being an end-goal? Despite the many obstacles outlined above with measurement currencies, envelopes of solutions and the political–scientific split, there is still scope for it being classified as an end-goal type of concept. However, for it to be a useful end-goal type of concept it would need to have a form from which starting points and processes could be designed. The initial trawl through the evidence in this chapter suggests this is not yet possible.

1.5   Taking the discussion forward

This chapter has therefore set the scene for the rest of the book. A brief look at the subject reveals a wide set of views each purporting to describe sustainable development as a concept. There are a number of important questions that arise when studying the subject and these will become constant themes throughout the rest of the book: Can useful theory and practice be developed without paralysing complexity? Does the concept lend itself to be starting-point, process or end-goal? How should development be balanced? A balance of least damage and most added benefit? Or one that takes account of future needs?

The initial study of the subject has been framed within the question of whether sustainable development can be classified as a starting point, a process or end-goal type of concept. A set of definitions and principles have been developed which could lend themselves to being a starting-point, but there are still difficulties with a common language.

The arguments for classifying sustainable development as a process-led concept have been briefly discussed. This suggested that balancing the social, economic and environmental concerns may be a misleading simplification, and that a common process appears to remain elusive.

The end-goal arguments have been briefly reviewed and highlighted the need for a common currency or proxy to allow measurement of success. It was noted that indicators represent the common tool at present. It was further noted that there is a need to acknowledge that any solution is part fact, part political decision.

There are various schools of thought on the subject, and the next three chapters set out to describe a set of major groupings of theories and definitions. As with all descriptions of such a wide and diverse field there is a large degree of subjectivity in choosing which theories to group into which school of thought. More importantly, the diversity of opinion within each school of thought may be such as to stretch the credibility of the initial reason for grouping in the first place. Thus, a degree of care is needed in the interpretation of this work.

Three groups have been chosen as representative of the range of views available in the field of sustainable development. The choice of group definition arises from study of a number of angles of thought, although it must be stressed that this is subjective in nature and others see more important defining features across the range of views:

(1) Mainstream Economist
   ■ Tends to relate to the status quo, unless evidence proves otherwise.
   ■ Basically believes that the current systems of choice, evaluation and decision, although not perfect, are the best available.
   ■ Often economics (with social implication) driven.
(2) Strong Environmental
   ■ Tends to look to stronger change as the solution.
   ■ Includes groups who believe current systems have completely failed mankind and the environment, and need complete removal and replacement.
   ■ Often environmentally driven.
(3) Middle Ground
   ■ Suggests need for some change.
   ■ Advocates of compromise who often believe that current systems need adjustment/change and more joint work.
   ■ Divides into distinctly different groups and promotes a range of drivers (technology, social, a balance of the two),

while some base their arguments on simple practical work.

The next three chapters therefore concentrate on the theories that exist and start by looking at the two outer edges of the envelope available: the traditional economic approach in Chapter 2 and the environmental response in Chapter 3.

2 The 'Traditional Economist's' View of the Debate

2.1 The basic arguments

The first chapter has examined the broad context of sustainable development. It was noted that there was no one fully accepted definition of sustainable development, and that there are at least three main dimensions to the subject: the social, the economic and the environmental. It was also noted that decisions on what constitutes sustainable development are partly political, partly evidence-based. The evidence base in turn is partly scientific and partly social evidence. Thus, decisions on the subject mix fact and belief.

The subject can therefore be studied from at least three viewpoints: the social, the economic and the environmental. In this chapter the views of economists will be studied – a view that has been very powerful for many years but which appears to be under increasing scrutiny. As evidence or the weight of opinion shifts the arguments show signs of change.

It would appear that the environmental effects and the socio-economic effects of economic decisions are key to the debate, and thus they will be closely examined in this chapter. The economists' view is an interesting one to start with because it claims to cover social factors within the theories, and increasingly also claims to cover the environmental issues.

Categorising all views in this simple manner risks causing offence, but increasingly the groupings are self-selecting (Appleyard 2001), even where the opinions within the group are very diverse or even blur into the other groupings. As with the other schools there are those who wish to retain the status quo (traditionalists) and there are those who want change (reformists). The chapter will consider initially the traditionalists before looking at the wider range of views within the school and the views from elsewhere. Thus, it must be made clear that this initial generalisation is for the purposes purely of discussing sustainable development in its widest sense.

It may be an injustice to economists to describe their position as being at one extreme of the range of interests since the beliefs have been mainstream to many development decisions for many years. However, in the interests of clarity it is useful to describe their position as being at one end of the spectrum and, additionally, to generalise about that position.

The economist-led approach appears to sit around the premise that growth, as measured primarily by Gross Domestic Product, is good. This is based on past and present evidence that prosperity in this form has:

(1) Paid for improvements in social, physical and psychological well-being
(2) That environmental mistakes have generally been rectified as prosperity increases
(3) That an economic-led progress plan represents the best (if still slightly flawed) approach
(4) That technological progress or invention solves all or most problems as they become acute and
(5) That fact- or evidence-based approaches are promoted through this approach.

In order to fully understand the economist angle to the sustainable development debate it is important to understand some of the basics that have helped to shape that viewpoint, and this is the objective of the rest of this section. It is thought that modern economics has at least a 500-year history. The rate of material output doubled between 1500 and 1700 and the speed of change was such that it became noticeable to the individual through his or her lifetime. The changes prompted Smith, Malthus and Marx amongst others to speculate on the hows and whys of economic growth, seeing the system developing as a mixture of sociology, politics and philosophy (Martin-Fagg 1996).

It is now widely accepted that economics is a subject that considers the issues associated with allocating resources in conditions of scarcity. Key to all of this are the concepts of supply, with its systems of production or service, and demand, with customer behaviour and choice. Without scarcity or where there is no exchange of product or service there is no economic market. A basic assumption is therefore that all resources, including time, are finite and need to be managed by society and individuals to satisfy needs (Mills *et al.* 1995). Economics provides interpretation and understanding of the system management, through creating supply and demand markets, evaluating them as they work or trying to balance the two sides. Economists therefore look for causal relationships in their analysis, but often with imprecise complex data.

In situations where there is scarcity and choice, prioritisation is important in order for reasoned decisions to be made, and it is widely acknowledged that economic theory influences politicians and political decision-making. Reasoned decision-making, it is believed (Mills *et al.* 1995), must involve reference to efficiency and optimisation of market conditions and the supply of product or services (note that all these concepts – politics, efficiency and optimisation – are considered again in Chapter 6). A key assumption here appears to be that markets function when there is an incentive to exchange goods or services, i.e., the exchange adds value to the supplier and consumer, and this contributes to wealth creation and economic growth.

An important factor for economists is the role of government in influencing the right conditions for supply and demand and their consequence on wealth creation from this added activity or from allowing freedom of choice. The choice is to leave supply and demand to sort itself out or decide that certain minimum standards of living must apply and markets must be changed if they do not fulfil those expectations. There is universal agreement on limiting the worst excesses of a free market (banning the worst monopolies, supporting the less able in choice or access, providing safety nets for those who fall out of the employment market, etc.).

It is therefore generally agreed that government has key roles in basic service provision, tax and spend activities, regulation to limit the dangerous or worst excesses and provision of fair access to opportunity, although debates continue as to the level at which state intervention should operate.

Mills *et al.* (1995) suggest there are five important issues for economists to consider at the national level:

(1) *Inflation* – a visible sign of a supply/demand imbalance or that costs have generally been set at the wrong level.
(2) *Environment* – where it is difficult to devise markets.
(3) *Unemployment* – a key visible sign of supply/demand imbalance in the social end of the economy.
(4) *International competitiveness of nations* – seen as an important sign of the health of a nation's supply/demand markets.
(5) *Supplier/consumer choice* – the degree to which a nation desires choice in its products and services (i.e., the breadth of the markets).

The evidence supporting these five factors is key to a politician's choice in setting the levels of government role, and is also key to believing that economics has provided an understanding of how the world works and continues to provide solutions. This evidence is critical to the economist's view of sustainable development.

The early part of the second half of the twentieth century brought prosperity to the Western developed world using the models and policy ideas of Keynes, and this appears to be a key historical period for modern economics. For most of this period, an inherent assumption within the theory, that growth was a part of the system with no limits, remained unchallenged as the evidence appeared to agree with this belief. Skidelsky (2000) suggests that Keynes' theory on supply and demand rested on a belief that the equilibrium between the two occurred at a point below full employment of the national workforce. The existence of money brought flexibility to the system of matching supply and demand, and became a store of value and the measure of wealth creation. However, it also brought the power to disturb the equilibrium, bringing cycles of over-supply or excessive demand. This could be mathematically modelled and used by governments to develop policy, actively manage the demand and supply within their economies, explain decision-making and develop controls to adjust the balance.

In developing such models in the late twentieth century many governments now emphasise adjustment rather than control, with balance attained over a cycle of supply and demand rather than at any given time. This reflects the reality of government's inability to either accurately predict all the activity within their nation or provide an accurate solution to issues which arise, i.e., it is a rough and ready system with many risks and uncertainties.

This element of instability led many to believe that, in fact, even adjustment was barely possible and therefore a market that was either free or with minimal control was the best model. When the former communist dictatorships of Eastern Europe finally began to unravel, it appeared that the model and free market policies had triumphed as the only solution, with no obvious alternative.

The belief in free market policies is not without its critics. It has been suggested (Middleton et al. 1993) that in order to counteract this belief and to provide an alternative to the harsher effects of unbridled free market systems the concept of sustainable development was born. Anecdotally, it has been suggested that the source was a group of left-leaning policy advisers to the likes of the World Bank and other global aid organisations who were seeing the effects of the free market on both environment and poverty in less-developed nations. Pearce *et al.* (1990) likewise suggest that the first stirrings of debate were about conservation and how this fits into an economic framework.

2.1.1   Some comment and analysis

Looking at the basics of the subject, it is striking to note the similarities in the principles and objectives set out in both economics and sustainable development. Both consider the fundamental problems of allocating (or not allocating) resources in conditions of scarcity. Economics seeks the opportunity to use, sustainable development seeks the opportunity not to use or to spread use. Both seek a balance or equilibrium for a system that is dynamic.

Skidelsky (2000) explains Keynes' philosophy of seeking systems that, for the best of reasons, promoted stability and civilising instincts, and were evidence based. Economics was viewed by Keynes as a study of logic and methods rather than objectives or outcomes. The indirect outcome was viewed as some form of well-being or 'good life', which again is similar to the many views described in Chapter 1, a similar outcome desired by advocates of sustainable development. Interestingly, Keynes noted that emotions were an important part of the system, i.e., the 'feel-good factor' is an important driver in humanity's economic behaviour, suggesting that money movement and wealth alone were insufficient to describe the system.

The question therefore arises of whether the two subjects of economics and sustainable development are in fact the same, merely looking at the evidence from different viewpoints (Fig. 2.1).

The sand-clock and the old argument of being half full or half empty?
Are we looking at the opportunities to use or the opportunities not to use

Fig. 2.1   Half full or half empty?

With the benefit of hindsight, it is striking to note the absence of any reference to the environment or the issue of resource depletion in much of the early work of economics, although the more general issue of scarcity was always critical (i.e., a scarce resource will command a higher price and thus the demand will be reduced until such times as an alternative arises). Thus, it seems logical that the environment would become an area of importance as scarcity occurred, while the social implication of economic systems continued to remain an open area of debate.

Initially, it appears that the concerns about the environmental and social implications of economics were practice-led and they lacked a clear explanation in traditional theory. It is important to note, however, that 'economics' likewise was a system before it was a theory.

Thus, sustainable development, which initially may have been set up as a challenge to the status quo within economics, may simply be a relatively young off-shoot of economic theory. This is a hypothesis that is examined again in Chapter 8. The basics of economics show similar principles and drivers to sustainable development and it is important therefore to further break down the sources of the economic viewpoint and the evidence. This is done in the following two sections by initially looking at the traditional arguments and then looking at advocates of reform to that traditional stance.

2.2   The traditional end of the argument

The well-publicised view of this side of sustainable development sees the supporters of a traditional economic-led approach to sustainable development centred around the world's biggest companies, the multi-nationals, and global financial institutions such as the World Trade Organisation (WTO), International Monetary Fund (IMF), Organisation for Economic Co-operation and Development (OECD) and the governments of the developed world (Appleyard 2001). Much debate revolves around the activities of the World Trade Organisation, which seeks to achieve consensus for a common set of rules on global trade policies across 135 countries. It has a broad agenda covering such issues as trade barriers, dispute settlements and specific sector agreements, and runs with the central tenet that free trade benefits all (Schott 2000).

In recent years, many of the objectives of these organisations have been challenged by advocates of stronger social and environmental policies. Adjustments to policies have sometimes occurred to make allowance for changing political perceptions or changes in the evidence, although the central assumption remains that the world can continue to grow ad infinitum.

The World Bank approach

The World Bank's position, for example, continues to advocate economic growth as a fundamental requirement, but it is couched in terms familiar to the sustainable development debate (Thomas 2000). The Bank, in fact, advocates a concept of 'sustainable growth', a concept that many would continue to query. The principles behind the Bank's policy are broadly in line with Brundtland's definition recognising the importance of physical, human (social), institutional and natural capital, all of which need investment and restoration. The Bank's priority appears to be addressing systematic failure and in particular the distortions of the market brought through badly focused subsidy of inappropriate activity. Three clear objectives emerge:

(1) The need to address inequalities in health and education investment.
(2) Tackling issues associated with poor governance which, it is believed, retard growth and damage the poor.
(3) The protection of natural resources, which it is believed is a strong indicator for a vibrant economy with social responsibility.

   The World Bank's position is subject to much close scrutiny and, as such, is continuing to evolve. Green non-governmental organisations have become a part of the advisory team with 70 secondees reported in 1999 (*Economist* 1999). The objectives, while often taken to be the global lead (and, as such, so openly misconstrued as poorly focused), are a reflection of the Bank's mandate to help the poorest countries to attain the standards of the developed nations rather than set the standard for the world.

Another useful source, which is used extensively here, is the *Economist* magazine which appears to represent a business school/ government treasury type of angle to the discussion. It is a standard bearer for mainstream economics, with a strong stance as a flag-standard for traditional economics, more traditional than the World Bank (*Economist* 2000a). Its use within this text is on the basis of three main factors:

(1) It is easy to read and understand – economists and economic theory can throw up some very difficult language which detracts from any strong basic arguments.
(2) There is good evidence produced in support of the arguments – articles in the magazine are typically accompanied by a well-sourced evidence basis.
(3) It is prepared to take a stand – as the standard bearer for convention the magazine is clearly willing to explain and argue its stance, where other standard bearers often change their view too quickly to be mapped.

Having looked at the principles in the last section, it is important to study the large body of evidence put forward by economists which shows the link between growth or wealth creation in their terms and improvements in social, physical environment and psychological well-being.

The traditional economics approach, as mapped by the likes of the *Economist* magazine, tends to have a set of implicit political assumptions in the background for reasons explained in earlier sections and Chapter 1. Thus, even the economists who stress the importance of an evidence base must still accept that part of their approach rests on political belief rather than pure fact.

The political line appears to support systems such as 'democratic liberal capitalism' (*Economist* 2000b), globalisation (*Economist* 2000c) and evidence-backed decision-making (*Economist* 2000d) where possible. The 'democratic liberal capitalism' advocated has an implicit small government argument that minimal intervention by government is best. The globalisation argument follows a line of argument that international competitiveness is the best direct measure of success and progress, noted previously with the list put forward by Mills *et al.* (1995). International barriers to trade tend to lock nations and their peoples into blocking progress (Streeten 2001).

The importance of the evidence-base approach is crucial to the economist's argument. Without evidence there is no model and without a model there is only uncertainty. This pervades all aspects of life. Modern risk management, for example, seeks to build mathematical models which lead directly to the development of new insurance markets, thereby reducing costs and improving the manner in which business is conducted.

Thus, economists believe that they have built a picture of an economic framework which explains and leads to progress in the socio-economic dimensions of sustainable development for the developed world (*Economist* 2000d). At a national or international level, this progress is measured through Gross Domestic Product (GDP), an imperfect but important hybrid model of the combined output of societies, which will be studied in more detail in the next section. There are few economists who see GDP as a perfect measure, but most believe that there are no acceptable alternatives at present, and many continue to see it as a useful proxy for wealth across both poor (developing) and rich (developed) nations (Martin-Fagg 1996; *Economist* 2000e). Many further believe that it will provide a framework to solve environmental issues if supported and advised with the right evidence-base (*Economist* 2000f, 2001a).

It has been stated that a vibrant business sector and economic growth, the cornerstones of the economist's argument, generate jobs, raise incomes and contribute to the efficiency of goods and services across the developed and developing world. The private sector, and multi-nationals in particular, has a prominent role to play by taking risks in less-developed countries, improving respect for law and promoting global standards of governance and efficiency. In Brazil, for example, a survey has shown that 50% of those surveyed who had been brought up in poverty were now above the poverty line. Income mobility, the ability to move from one level of income to another, has become widespread, and while absolute numbers have not changed, the percentages have improved as the global population increases. The conclusion is that growth of the private sector is good for the poor (Pfefferman 2001).

Health is a primary factor in the well-being of humanity and there are numerous studies to indicate how health has improved throughout the ages as wealth has increased. Child mortality rates, life expectancy and other healthy living indices all show improvement in most developed countries where wealth is also growing. Economists suggest that higher prosperity allows higher health-care spending which, in turn, results in better health (*Economist* 2000g). It is noted, however, that the effect is indirect.

Likewise, poverty alleviation (*Economist* 2000h) shows improvements for most of the population as wealth increases. Many economists believe in the 'trickle down effect', i.e., that new wealth generally created by a few very quickly translates into reduction of poverty across the board (*Economist* 2000a). It is clear, however, that problems arise with this particular aspect of development because there are a wide range of definitions used for poverty, which provide differing interpretations of the results. It is, for example, acknowledged by the World Bank that the number of poor in the world has increased by 100 million in the decade 1990–2000. Thus, the World Bank has switched its view from *all* economic growth is good to *better* economic growth helps to reduce poverty. The Bank now links investment in education, investment in health and attitudes to environmental protection and governance, and sees these factors as being important to the definition of sustainable growth with benefits to all (Thomas 2000).

An interesting factor often quoted by economists as a vital part of sustainable growth is foreign direct investment, the amount of foreign capital funding feeding into a country or region. Although the link between this and sustainable development is hard to immediately identify, it is viewed as a measure of the world's confidence in the country of study, has direct consequences for increasing prosperity and employment, is seen as a force for 'change' (an important subject which will be studied in more detail in Chapter 7) and has also indirectly been linked to improving transparency and reducing corruption in business and policy implementation (*Economist* 2001b). These are viewed as important factors in the social and economic dimensions of sustainable development. It is, however, a politically sensitive subject since it is also associated with loss of local decision-making and a promotion of centralised corporate power concentration, both important to many sustainable development enthusiasts who advocate the principle of equity.

Much of the success associated with foreign direct investment is generally concentrated on rich nations becoming more successful, although one noted exception is China (Fig. 2.2). There are suggestions that it has a direct effect on increasing the gap with poorer nations, both in terms of wealth and access to decision-making.

Productivity is another indirect factor promoted by economists as a positive in the drive for sustainable development. Increases in productivity mean increases in efficiency, which in turn promotes better use of resources. In practice, however, productivity, as measured at national levels, is as much affected by currency changes and the manner in which it is measured as it is by any true efficiency gains (*Economist* 2000i).

Technology is seen as an important driver in meeting the challenges posed by sustainable development. The latest advances in computing, telecommunications and the pure sciences have the

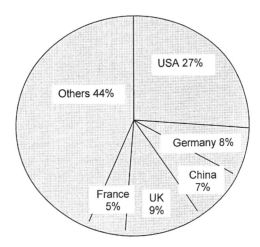

Fig. 2.2 Foreign Direct Investment – the main recipients. Source of information: Economst 2001b.

potential to raise performance and improve efficiency. The trickle-down theory again features prominently in this argument, suggesting that it will benefit all parts of society including the poor and address the problems of the environment. It is clearly assumed that technological advance, just as with economic growth, is a natural assumption within the system (*Economist* 2001c).

In the same manner, economists appear to believe that the environmental dimension to sustainable development can be reduced to just another market sector problem, which needs to be isolated, analysed and solved.

Environmental issues, which are given greater prominence in the sustainable development than the general economic debate, are viewed as yet another example of how general wealth increase brings about improvement. The case study in Section 1.4 which highlighted how unsustainable behaviour in developed-world cities is often accompanied by a later corrective action leads many experts to an optimistic belief that all or most damage is repairable.

Economists point to a damning lack of credible environmental evidence as a major problem in addressing environmental concerns properly or efficiently. A number of all-encompassing cross-disciplinary studies have been conducted across the globe, but many fail as a result of poor-quality environmental data and the complexities of their interpretation.

One study, which appears to have the approval of mainstream economists, is the Environmental Sustainability Index (ESI) which attempts to classify countries using 67 separate variables. Indeed, an explanation of the ESI is presented in the *Economist* (2001a) in terms of its value in bringing order to the previously unordered field of environmental data analysis and its attempts to link 'prosperity and greenery'. Thus prosperity measured through GDP remains the core objective, with environmental issues a credible but secondary concern. Although the ESI is an imprecise tool it is worth noting since it has allowed economists to review their case from an apparent environmental angle.

The ESI study in 2001 (Table 2.1) had five broad categories for national comparisons: environmental systems health, environmental stress caused by human impact, the vulnerability of local human systems to environmental disruption, social capacity and a country's record on dealing with global issues. The results indicate that environmental sustainability as measured through these five broad categories is highly correlated to per capita income, i.e., the richer a country becomes, the more it can afford to spend on the environmental debate, promoting the simplistic belief that economic growth is beneficial to the environment. An interesting factor was the perceived link between corruption and its detrimental effect on environmental sustainability. Again, the belief is that a market working properly will have respect for all things including the environment.

Table 2.1   The top and bottom six countries in the ESI study (Esty *et al.* 2001).

1. Finland	117.	Nigeria
2. Norway	118.	Libya
3. Canada	119.	Ethiopia
4. Sweden	120.	Barundi
5. Switzerland	121.	Saudi Arabia
6. New Zealand	122.	Haiti

An interesting facet of the *Economist*'s approach to sustainable development is its view within the debates on global warming. Initially the belief was that without hard evidence there was no point in dealing with the issue as a problem. By 2000, and long after most players had accepted that global warming was occurring, there was a recognition that it exists but could only be properly dealt with when proper evidence was available (*Economist* 2000j).

Environmentalists would rightly point out that warning signs were there even if the evidence was poor quality and prevention was a better option. The economists in turn suggest that while the science was uncertain, the costs of taking preventative action which may not be needed was a waste. Further, they believe that more information leads to a better focused solution, that new technology has in the past and will in the future solve the problems and that increased wealth brings increased funding available to pay for clean-up or problem-solving (*Economist* 2000e).

On global climate change, there is now a realisation that the problem may be accelerating, with even the economists accepting that, despite the uncertain science, preventative action may be necessary. The Kyoto Convention of 1997 committed developed countries to action to reduce emission of greenhouse gases, and was followed by the Hague Convention in 2000 which was, in theory, designed to seal the deal. Interestingly, the solutions advocated by traditional economics revolve around tinkering with market mechanisms to allow the private sector to deal with it in an efficient manner. There is some thought given to compliance and sanctioning for worst offenders but a desire to steer away from much proactive action by the public sector or governments at a global level (*Economist* 2000e).

The initial resistance to action on climate change from traditional economists has been chiefly on the basis of the cost of making environmental improvements, mainly in developed countries. However, even economists themselves are starting to query the estimated costs since analyses vary from net positive gain for humanity through to monumental loss. The UN's Inter-governmental Panel on Climate Change has estimated the cost of change to be between 0.1 and 1.1% of GDP in 2010, the year of implementation of action arising from Kyoto (*Economist* 2001d).

There is a strong view in some developed-world governments that developing nations have an equal obligation to deal quickly with the environment and climate change. The idea of trading emission credits between developing and developed nations is one possibility to involve both sets of countries, although some developing countries such as Argentina and Kazakhstan have gone one stage further and already announced voluntary national targets (Dunne 2001).

Even in the admission that climate change may be happening there is still a belief that solutions being put forward could damage the economy (*Economist* 2001c), that economic concerns are paramount and that decisions on any other basis are unworkable or unfair. Despite this, the traditionalists now appear to accept that evidence points to damage occurring in some aspects of the environment. The most important changes in belief have been the growing acceptance that global warming and the greenhouse effect are occurring and that the cause of the damage is material consumption, a direct output of capitalist systems (*Economist* 2001e). The answer of traditionalists to this problem is, of course, market-based in their application; through creating new markets in pollution and clean-up, developing new forests or a small degree of market control or tax incentives.

Market-based sustainable development

An example of this is the sustainable management of rainforests in Brazil (*Economist* 2001e). Deforestation of the Amazon jungle has been as much an economic disaster as an environmental one. It has been argued that the gains made in cutting down trees, exporting them and using the land for agricultural use have been vastly overestimated (i.e., the cost-benefit analysis would never have stacked up) and the system of grants attached to this change of land use was perverse and open to misuse.

Once the system of grants was reduced or removed in the mid-1990s, good quality forested land was worth 40% more than cleared land. The introduction of satellite imagery has allowed law enforcement agencies to better monitor illegal activity such as land grabs or illegal felling, and there are now certification schemes for well-managed timber for export markets. In recent years, timber companies have come to realise the benefit of unplanned felling and have started to 'manage' a 30-year rotation cropping system. To add to this, the oldest specimens are left untouched. It is claimed that this allows conservation schemes for trees, plants and animal life to run in parallel with the commercial felling activity.

Interestingly, it is noted that one of the biggest future pressures on the successful continuation of such schemes is population pressure, since the area's population is increasing by 3.7% a year, prompting a search for 'sustainable living' schemes to run in parallel with the sustainable management of the forest. The net result is that the Amazon forest, which in the early 1980s was predicted to disappear within 20 years, now has a predicted life of 200 years.

Indeed, a general rule of the more traditional arm of economics is that it is important not to assume that one sector should not have greater priority over others or be seen to be more essential than others (*Economist* 2000b). This effectively rules out special treatment of the environmental sector.

2.3   The reforming end of the debate

It is clear, however, that attitudes are changing as evidence improves. The European Union (EU) increasingly views environmental concerns as an important parameter in decision-making and has taken many measures to incorporate environmental protection into legislation (*Economist* 2001d). Pearce *et al.* (1990), in work for the UK government, acknowledged that the issues of environment, futurity and equity were not fully served in economic theory. These are key principles for many in the sustainable debate, as shown in Chapter 1.

Thus, there are many who continue to support the belief that the current economic-based system is the best way forward but who would point to many situations where free market policies and models fail. They suggest the need for fundamental reform rather than tinkering at the edges or specific action on a few subjects. Hawken *et al.* (1999) provide a particularly clear explanation of many of the reforms needed and Pearce *et al.* (1990) provide good supporting evidence for why the reforms are necessary. The basic premise of this reformist-type school of economics is that indiscriminate unmanaged economic growth harms social and environmental systems because, as even traditional economists will agree, the system model is imperfect. The imperfections appear to be consistently skewed towards more damage in social and environmental systems.

In summary, Hawken *et al.* (1999) provide a useful list of the defects of current economic theory and policy:

- Lack of social dimension, fairness dimension and the belief in self-policing in the current political stance which supports the theory. Markets need oversight or supervision in order to make them 'civilised' and there is a range of choice between free and heavily regulated. The current choice promotes free market but has its down sides in areas of human behaviour where the theory has little impact.
- The assumption of a perfect market and rational behaviour which underpin the theory. In fact, they list 18 important

assumptions which are incorrect in free market economic theory.

■ The effects of subsidy, market failure and the speed and flow of capital in today's system, i.e., the inefficiencies that form part of any imperfect system and which are multiplied, exploited and replicated through a quick-moving system.

■ The priorities established within the system, such as lowest initial cost as the basis for procurement decisions. They suggest that this factor is significant in causing serious misallocation of resource, inappropriate prioritisation of future development and incorrect attitudes to risk and research.

■ The bias against long-term decision-making which arises from a reliance on discount rate methods, a factor that is acknowledged by many even in the traditionalist camp.

The perfect free market assumes provision of perfect information, perfect competition, no regulatory imperfections and with all risk properly calculated. However, as Hawken *et al.* (1999) point out, such a perfect market would produce only average profit for all, and all incentive to improve or trade would disappear, so clearly the market in practice is very different from the market in theory. Monopolies and other 'unfair' play, misallocation of capital, organisational failures, regulatory failures, informational failures, value-chain risks, false price signals and incomplete markets all contribute to the imperfections.

Market failure – energy efficiency

A useful example is the case of energy efficiency. For the business sector, energy savings are operating cost cuts which are added straight to profit. They should therefore be highly significant and attending to them should be rewarding to the company. However, it is often viewed as low priority and seldom reaches decision-making agendas since, for the average business, it represents a small percentage of costs (typically 1% for a commercial office-based business). Thus capital is misallocated and the best return on investment is not chosen because it is misperceived to be too small. The decision to buy new energy-efficient equipment is frequently subject to the lowest capital cost or short-term payback calculations when, in fact, it is a long-term investment with continued payback over 15–20 years in many cases. Thus the tools for investment are inadequate.

In the home, rational behaviour is even less in evidence. Hawken *et al.* (1999) explain how the use or not of an air conditioner is the result of household schedules, misperceptions, noise, health and attitudes to machinery. Only 25–35% make a decision based mostly on costs, as the economic theory of behaviour would predict.

The environment suffers from its free availability in many cases. Items such as landscape views, natural water purification or the productivity of mangrove swamps have no directly measurable economic value and thus are accorded a zero price. No market exists in them as a commodity and they become free resources. They are thus exploited, as economic theory predicts, with no constraint until capacity is exhausted.

Another important factor, futurity, is explained by Pearce *et al.* (1990) as extending the time horizon of development and planning to the longer-span future inherited by future generations. From an economic point of view, a key factor is the argument between anticipatory and reactive policy and the point at which evidence becomes accepted that damage has occurred and is irreversible when the information is not initially clear-cut. This concept of evidence threshold is an important one which is raised again in the next chapter.

Equity, fair treatment of the least advantaged and future generations, is seen by Pearce *et al.* (1990) as a goal of economic policy and must be an important addition to the best economic development frameworks. Pearce *et al.* (1990) suggest that pollution at a global level is the biggest priority, sitting as it does at a level beyond current national and international policy frameworks, often beyond regulatory or market controls. Many in the traditional camp believe that any willingness to take action to protect the environment and ensure equity is directly related to the wealth of a nation or region. Reformers would argue that this is not the case, as Fig. 2.3 shows. This implies that it is actually a choice made with greater reference to necessity than wealth. In effect, sustainable development is not a bored, rich man's cause but rather a fundamental concern of the global population.

Pearce *et al.* (1990), writing for the reform camp, suggest that man-made capital is not and can never fully be a substitute for natural capital. Issues such as natural resilience, eco-system stability and natural carrying capacity or productivity do not wait for the wealth of a nation to reach the point where they can be addressed. Reform-minded supporters of the economic system

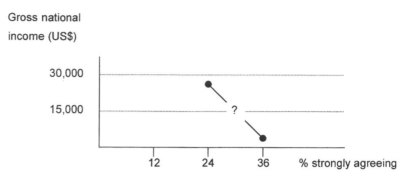

Fig. 2.3 Willingness to pay 10% more for the environment, based on a sample of 12 countries which were then grouped into either OECD high income or non-OECD low or middle income countries (*Economist* 2001a).

therefore suggest the need for more vigilance, setting clear goals which incorporate social and environmental concerns, identifying dysfunctions in the market for removal or improvement. They acknowledge the basic market system as providing the principles to cost all goods properly and the opportunity to create new markets armed with new information. This is in contrast to many of the approaches highlighted in the next chapter which would suggest that the system has too many faults for reform and needs to be replaced.

2.4   A critique and analysis

Having looked at the traditional and reform ends of the economics school it is useful to review and analyse the implications for this on the arguments from the first chapter. Some of this analysis has already taken place in the early section which looked at context. This concluded that the principles did not vary greatly since both seek to improve the quality of life and to study the allocation or non-allocation of resources while acknowledging scarcity. This section therefore concentrates on the implications of the evidence and outcomes that economists seek.

The traditional economic-based approach relies on evidence. It appears, in a sense, to come with an optimistic assumption that current systems are positive (and innocent) until proven otherwise. This is because economists believe that they have proved that the system works in general. By contrast, the 'opposition' views, some of which are detailed in the next chapter, hinge around preventative principles, i.e., evidence will come when it is too late to improve or amend the situation and as such damage must be avoided before it happens. Thus, in a sense, it assumes guilty until proven innocent, the exact opposite.

Both views therefore have a strong basis in logic and, in fact, in legal circles there are similar arguments, e.g., is it better to prevent crime or should assumption of innocence prevail? The big question therefore remains of how robust is the evidence in general and, specifically, of growth being the only factor that supports improvements in social, economic and environmental welfare. Is that evidence strong enough to counter the claims of environmentalists and social scientists that it is one-sided and self-fulfilling?

All of the previous arguments rely, to a greater or lesser extent, on evidence based on the basic measure of growth and prosperity used by the economists, Gross Domestic Product (GDP). Martin-Fagg (1996) provides a useful introduction to the basics of GDP and its measurement. It can be measured in a number of ways although there are two main methods at national level (Fig. 2.4).

It is often stressed that this is an estimate and there are many adjustments made to the figures. It is income- rather than stock- or asset-based and in this sense is not a direct measure of wealth. It does not give a full picture of the economy, since activity that does not involve exchange of cash is not covered. Housewife activity, for example, does not feature because it is an unpaid activity and therefore it is believed that GDP is generally an underestimate of economic activity in most cases.

As well as omitting many useful activities, critics would point to many activities included in GDP figures that do not add value to the economy or which are clearly negative in impact. Many aspects of GDP calculation actively promote detrimental behaviour in social and environmental spheres of life, interpreting them as positives adding to GDP. An oft-quoted example is the effects of a road accident or an oil spill which, because of the activity associated with the clean-up, would be included as adding value to the economy although it clearly does not add value to the quality of life. Another important factor missing would be the depletion of natural stock, which again has a strong long-term effect on a nation's wealth (Meadows *et al.* 1992). Hopwood (1999) has reconstructed the UK 1999 figures for GDP by classifying them in terms of activities that add capital stock, administration costs, those that contribute to the satisfaction of human needs and wasteful activity. The result, Fig. 2.5, shows the sizeable percentage

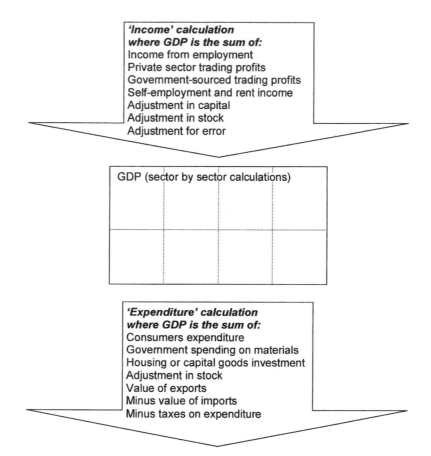

Fig. 2.4 Calculation of GDP.

of activity that is classified as adding no positive value (waste and administration).

Thus, GDP as a working tool is flawed, although many economists would argue that it is still the best measure of growth or progress available. Pearce *et al.* (1990) in their review provide a useful summary of the possible alternatives to GDP. They acknowledge that GDP/GNP measures of development do not satisfy sustainable development principles and do not take proper account of environmental concerns. They provide an excellent review of monetary methods of valuing the environment and physical (land use, energy, emission accounting) methods but conclude that all have sufficient concerns that they can only augment, rather than replace, the current measure. Sustainable development indicators, to round out economic valuing, are put forward as a possible way forward.

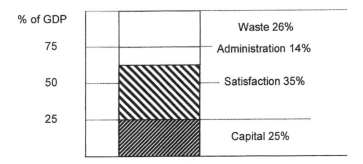

Fig. 2.5   The 'good' and 'bad' in GDP (Hopwood 1999).

Another interesting approach is through the adjustment of the calculation of GDP, to avoid the most obvious flaws and improve the coverage of missing elements such as environmental damage. Daly and Cobb (1989) pioneered an index called the Index of Sustainable Economic Welfare (ISEW) which has received world-wide attention. The basic method takes the GDP of a region or country, calculated through personal consumer expenditure, and adds costs or benefits for a number of additional factors:

■ Environmental degradation
■ Depletion of natural resources
■ Unpaid labour such as domestic housework
■ Social costs associated with an imperfect market

A study of the UK economy using the index (Jackson *et al.* 1997) suggested that while GDP increased by 44% between 1976 and 1996, the ISEW declined by 25% in the same period. This would suggest that improvements in economic wealth were at the expense of social and environmental degradation. However, it must be noted that the approach is relatively new and the basis of costing social and environmental effects needs considerable testing before it will be acceptable to a wider audience.

The clearest arguments in favour of the economist approaches are the constantly stated belief that, no matter how imperfect, past evidence suggests that it has worked before, and current trends suggest that it should continue to work for the foreseeable future.

In the socio-economic sphere of influence, viewed as an area of relatively sound evidential support by the economists, a number of serious questions evolve about economics framework and their deficiencies. Puttnam, for example, (*Economist* 2000k) suggests that the tenuous link claimed between social and economic wealth has been broken, and while physical and human capital have grown in recent years in the US, social capital has fallen. His view of social capital is based on 'informal social connectedness and formal civic engagement', and his studies suggest that society is leading towards social isolation. In the UK record numbers of houses are being built for single occupancy, reinforcing that viewpoint. From this angle, quality of life, an important factor in sustainable development, is not improving despite improved prosperity. Puttnam's view suggests that economic measures may be too narrow to cover just the wider implications of quality of socio-economic life, even before environmental concerns have been considered.

However, even the basic dependence on 'evidence' and the type of evidence is often open to question. There are the inevitable problems with methodology and assumption inaccuracies (*Economist* 2000l). It is often the case that the evidence arising from economic modelling is not clear-cut, which often leads to long time-lags before evidence can be validated to an acceptable degree (*Economist* 2000m). There is also the issue of difficult-to-measure secondary side-effects from actions taken to promote economic development (*Economist* 2000n). These all raise queries about the strength of the evidence, the selectivity involved in choosing the evidence and the methodology of analysing it.

A good example of this is the ESI case study mentioned earlier in the chapter. Although a very interesting analysis, there are many other such studies in sustainable development and one must query why this particular model has gained great credence with the traditional school of economics. Is it because it has come out of the same north-eastern USA business school fraternity that seems to foster World Bank/United Nations/US government policy and that it provides the answers that these organisations seek – a self-fulfilling solution?

The difficulties of measuring productivity

A good example of the lack of clarity from some of the evidence is the debate that has surrounded the belief that Information and Communications Technology (ICT) will have a significant effect on a country's productivity. A long-running dispute between leading economists in the US has shown how methodology, the assumptions behind models and inaccurate and complex data combine to produce a wide range of views on the subject. By the year 2000, some were claiming widespread productivity growth, others claim it was focused purely on the ICT sector itself, others had still to see evidence of productivity growth (*Economist* 2000j).

One factor, which indirectly affects the economists' position rather than being directly critical of their arguments, is the current tendency of some of the biggest and strongest supporters to change direction or, at best, appear ambiguous in the debate (*Economist* 2000a, Thomas 2000). The traditionalists have noted how bodies such as the World Bank and multi-nationals are increasingly changing policy in line with the desires of environmental and social pressure groups, and governments are offering apologies rather than explanation for globalisation. This suggests, incorrectly, that the arguments are weak when, in fact, it is more likely to be the politics of seeking a quiet life which prompts reversals of policy.

Diversity presents another set of problems. Thus, the behaviour being mapped seldom produces a consistent range of evidence. Cultures vary (Berry and Houston 1993) bringing with them different attitudes to accepting rules, ethics (*Economist* 2000o) and differing levels of desire for capitalism/state intervention. Countries sit at different levels of development, bringing different priorities, while at community level, Mills *et al.* (1995) suggest that step changes in social behaviour are often more significant than the extrapolated gradual change normally assumed in modelling.

2.5 Summary

The advocates of retention of the current economic system as the prime framework in progressing humanity and its environment put forward many good arguments. They rely on a system that has brought many benefits in the past and continues to provide a good though imperfect model for current systems. They acknowledge its imperfections and suggest new additions, amendments or some reform to improve the framework. Generally, however, they are in favour of the economic status quo and the central measure of GDP as the proxy for development.

Although slated as having the worst of objectives, the market systems brought forward by economists are based on theories with good intentions. Evidence still supports many of their assumptions, although it is not overwhelming, as evidenced by the healthy debate within the profession itself. It is becoming clear, however, that a few of the gaps in the theory are significant; areas such as the environment and futurity present problems which may need major adjustments of the current system. The implications of this will be further discussed in Chapters 3 and 7.

The basic assumptions of 'sustainable growth' and 'sustainable management' of the environment appear to be the two main components of the definition of sustainable development in the eyes of traditional mainstream economists. There are clearly many questions raised by the economist's arguments in the debate, but are there enough to call for a 'scrap and start again' policy? The traditionalists see the need for change as an opportunity for market development but without change of the fundamentals.

It is interesting to note the views of the reformers, however, who acknowledge the failures in the current system. Despite the problems, reformers generally favour the current systems, although they recommend adjustments. Hawken *et al.* (1999) suggest that basic improvements would rely on a better approach to design and the incorporation of design so that there was better generation of ideas, longer-term thinking and better, quicker identification of market failure.

Pearce *et al.* (1990) come out strongly in favour of the current system despite the reservations about valuing the environment, futurity and equity. They suggest a number of improvements, with the three main adjustments of current economic systems to more fully account for environmental issues, compensation policies to deal with futurity and ultimately better economic growth (like the World Bank) to promote equity.

There are many outside of economics, however, who would suggest that it is now recognised that such narrow definitions of well-being as GDP do not cover all of what is important to mankind and its environment. Other factors such as the spread of wealth, the quality of life/well-being, the environmental resource efficiency and perhaps waste minimisation (which feature in the next two chapters) all need due consideration.

3 The 'Environmental' Arguments

The period 1960–2000 witnessed a growing awareness of environmental problems across the globe. Issues such as local pollution, resource depletion, global pollution and loss of animal species gained prominence and intensified media interest despite, as stated earlier, a move towards remedial action to alleviate or remove the worst environmental problems in the developed countries of the world. The acceleration of growth and the speed of change that accompanied globalisation have prompted many to worry that an environmental disaster may be just around the corner.

Despite the legitimate concerns of environmentalists, many in the popular media (*Economist* 2000a, Appleyard 2001, Murray 2001) portray a growing gap between the views of mainstream economists and a disparate bunch of angry opponents which include anti-capitalists, green politicians and ecologists, many of whom are labelled as 'environmental'.

Before studying the various political and science-based standpoints, the initial focus will be on the environmental evidence that has pushed the debate along, factors that have united all shades of environmentalists. Many in the green movement point to existing economic systems, the unfettered desire to control nature and the leaving of markets free without looking at the consequences as the main faults that need remedy through a policy of sustainable development.

The events of Seattle, Genoa, Stockholm and other apparent street clashes between the two extremes are ignored in this chapter in favour of a review of the rational arguments, the players and the evidence presented around the environmental aspect of sustainable development. It is, however, inevitable that some reference must be made to the clash of interests and Section 3.3 therefore looks at how traditional economists perceive their opponents in this area.

Just as there is no clear single view from economists on the subject of sustainable development there is a similar diversity of views across the environmental lobby. The views range from science-based approaches through to politically based stances. The range covers those supporting a slightly amended status quo (such as the reform end of economics noted in the last chapter) through to the 'stronger' green fraternity seeking a completely alternative vision of development.

Economic growth as a basic assumption of society is questioned by some and accepted by others. All the players agree that its effect on the environment needs further study, an interesting alternative to the *Economist* strand of economics which views growth as purely positive.

The environment has always been an issue of interest to humankind. From primitive times, when man was a simple part of nature, there has been a desire to control and master nature in order to (1) avoid the worst excesses of it and (2) harness it for the good of humanity. This desire to control led humans to see themselves as the centre or pinnacle of nature. Two quotes from the Bible (*The Holy Bible* 1995) show the long history of this belief in western society:

> Genesis 1/28:
> '... replenish the earth and subdue it: and have dominion over....'

> Psalms 8:
> 'For thou has made him a little lower than the angels ... thou has put all things under his feet....'

Meadows *et al.* (1972) raised the debate of whether continued growth with the economic policies of the day would take humanity to the limit of the earth's resources. Their work looked at trends in population, food production, pollution, etc., and noted how many were exponential and showed the dangers of ignoring limits. Moreover, it has already been noted that the modern concept of sustainable development arose from a strengthening in arguments around nature conservation in the 1980s (Pearce *et al.* 1990).

The World Wildlife Fund perspective

The World Wildlife Fund (WWF) has often been put forward as a major source of inspiration on the green side of sustainable development. In 1980 the WWF was involved in the production of the *World Conservation Strategy* (IUCN *et al.* 1980). This had three main objectives:

(1) Maintain essential ecological processes and life-support systems
(2) Preserve genetic diversity
(3) Ensure the sustainable utilisation of species and ecosystems

All three objectives supported an environment that ultimately would benefit humanity from a scientific stand-point. The report, however, also argued for a number of policy changes, necessary if the above were to be achieved.

In 1991 the WWF was involved in the production of a follow-up report *Caring for the Earth* (IUCN *et al.* 1991). It had a definition of sustainable development as 'improving the quality of life within the carrying capacity of supporting ecosystems', and recognised the widening of the concept to include economic and social issues, although with an emphasis still on conservation.

The organisation has always stressed the importance of conservation, but its views have moved with the times as the sustainable development debate itself has moved (WWF 2000). Ideas such as limiting arms production to release budgets for more positive expenditure, the importance of personal attitudes, the link between poverty and environmental policy and the need for partnership have all been put forward in the search for better development.

It is acknowledged by all players that some existing development methods have in the past damaged the environment. Many thought there was a link between economically driven and measured development, and negative impacts on biodiversity, ecosystems and the build-up of pollutants led to concern about the sustainability of nature. Environmentalists suggested that it must be given greater precedence in decisions. Many policy- and decision-makers feared, however, that decisions based purely on environmental concerns were likely to harm mankind and undo the economic development that had served the purpose of mankind in the past. This set the scene for the initial ambiguity described, for example, in the Brundtland report which acknowledged the need for both economic growth and protection of the environment (Brundtland 1987).

As noted in the first chapter, some of the claims of environmental damage are accompanied by strong evidence, some of the evidence of the extent of damage remains disputed, while some of the damage has been shown to be reversible. The next section looks at the types of damage and some of the evidence produced to support the arguments.

3.1   Environmental evidence

In Chapter 2 it was postulated that the nature of any evidence, the source and its robustness are as important as the conclusions that can be drawn from it. The evidence that economists can put forward has many drawbacks because it often only shows indirect links between cause and effect, it frequently depends on the definition of the methodology or the type of results desired and it tends to rely on human behaviour which is fickle. The evidence available to environmentalists has many similar problems. It does, however, appear to have one advantage over the economist's evidence in that much of it is based on scientific fact, although that assumption will be queried later in the chapter.

This section has been split into three parts. The first part studies evidence where all parties now appear to agree that development has caused damage to the environment, the second part examines areas where there is still a debate on the extent of damage and the third part looks at areas where damage has occurred but it can be reversed.

3.1.1   Unquestionable damage from man's development

There is strong evidence that a number of factors are damaging (or 'changing' in the language of some of the players) the environment. The biggest priority in the eyes of many appears to be climate change, although there are equally serious effects through overuse of materials or the build-up of toxins in the food chain and the water cycle. Industrialised society appears to be a culprit in many of these. Many environmentalists would like to see industry pay the price for this failure through claim and bankruptcy. Other players, however, see industrialists as a key to developing more sustainable alternatives and would prefer to give them the latitude to come up with viable alternatives. Much of the argument depends on the type and strength of evidence available.

Climate change

The Kyoto Pact of 1997 set goals for reducing emissions of carbon dioxide, although many key aspects of the agreement have been challenged by key players. The US government rejection of the

Kyoto Pact goals in 2001 queried both science and objectives, although increasingly the scientific evidence is accepted by both critics and advocates of the Pact. Many believe the science behind the evidence to be inexact, but it is still believed that economic loss is resulting. The insurance industry, for example, reports a huge increase in natural catastrophes in the period 1990–2000 (Dunne 2001). In the two decades previous to 1990, yearly costs of insured losses from natural disasters were generally less than $5 billion (2000 prices). In the period 1990–2000 this rose to an average of roughly $15 billion annually.

Without heat being trapped by the earth's upper atmosphere it is believed that the earth's surface would be $-20°C$. Thus, there has always been a natural trapping of heat by this upper atmosphere, although the last century has seen an acceleration of the rate of heat being captured. A recent modelling of this (*Economist* 2001), which has drawn much support, suggests that man-made gases (the so-called greenhouses gases such as carbon dioxide, ozone and methane) are particularly good at trapping long-wave radiation and thus retaining heat within the atmosphere. To make the modelling accurate the effects of man-made gases have to be sifted out of the naturally occurring heat-trapping that occurs. The United Nation's Intergovernmental Panel on Climate Change (IPCC) reports that, in the twentieth century, the planet's climate heated up by between 1.4 and 5.8°C. It further reports that most of the warming over the period 1950–2000 was due to man-made phenomena although both data and modelling are still imprecise (Dunne 2001).

The possible consequences of further climate change may be that the earth's surface temperature may rise by a further 2–3°C over the next century. This may result in changes to global climate such as more desertification and more storms, some warming of seas and the melting of polar ice, which will lead to rises in sea level. It has been estimated that $5,000 billion of problems will result if no action is taken, although most of this will occur in the developing world (Lomberg 2001). Low-lying countries are likely to be submerged in some cases. Middleton *et al.* (1993) have listed the ten lowest countries of the world, all of whom will have significant areas submerged if current trends continue: Bangladesh, Egypt, Gambia, Indonesia, the Maldives, Mozambique, Pakistan, Senegal, Surinam and Thailand.

At the other end of the spectrum, the USA produces 25% of the world's carbon dioxide emissions (Dunne 2001) and any policy change designed to reduce emissions is likely to most affect its businesses. Table 3.1 shows the vast differential in wealth and population between the USA and the ten lowest countries. Taking action is likely to be costly to the USA and other developed countries and would have an impact on global wealth creation. However, no action may ultimately affect significantly more population. It has been estimated that the cost of implementation of the Kyoto Agreement for the USA, EU, Japan, Canada, Australia and New Zealand would be $346 billion by 2010 (Lomberg 2001), significantly less than the global loss of Gross National Income from ten drowned countries.

Table 3.1   The cost of climate change effects (World Bank 2001).

	Gross National Income (1999, $ billion)	Population affected (million)	Cost from Kyoto
USA	8879	278	Implementation
Ten lowest countries	452	620	Non-implementation

Who pays and who suffers is a classic anticipatory versus reactive policy dilemma. On one side, a moral argument supports the anticipatory policy of imposing emission limits, on the basis that it will devastate a larger though poorer population income. On the other side, supporting the reactive policy is the straight cost comparison as measured by Gross National Income above and an indirect argument which assumes increased future wealth will contribute to helping people move before they and their environment are damaged or changed.

Loss of biodiversity

It is widely believed that the loss of biodiversity is accelerating. This belief, however, hides the problem that it is not known how many species there are in the world; 1.7 million species have been identified from a possible total of 14 to 100 million estimated in existence. Some studies have suggested, for example, that 34% of fish species and 25% of mammals are currently threatened with extinction. Again, Lomberg (2001) produces different figures suggesting that fewer than 1% of current species will become extinct over the next 50 years, although this is still well ahead of the historical trend.

A major source of concern adding to this problem and to global warming is loss of tropical forest, since it is believed that half of the world's species reside in the tropical moist forests which occupy 6% of the world's land area, much of it in the developing world (Middleton *et al.* 1993). In developing countries such as the Philippines and El Salvador the rate of deforestation was over 3% annually during the period 1990–1995, although some reforestation is occurring in developed countries and, globally overall, forest cover has increased (Smith 1999). Again, however, the picture is complicated because new forest plantations are very often single-species, commercial ventures such that reforestation does not necessarily contribute to conservation of biodiversity.

Pollution

Pollution is a major concern at both local and global level. Agricultural processes, industrial processes and domestic human activity all contribute to the problem. Middleton *et al.* (1993) report, for example, that 17% of the world's vegetated land had been degraded between 1945 and 1990, with the worst region being Central America where the figure rises to 25%. Ninety-seven tonnes of fertiliser per hectare is used on average across the world together with large amounts of pesticide, often leading to pollution of water courses. It is reported that Calcutta alone pumped 400 million tonnes of raw sewage annually into the rivers which flow into the Bay of Bengal. The net result of agricultural and human waste is huge increases in nitrogen and phosphorus, often leading to algal blooms, a health hazard for many species.

Industrial pollution has caused all manner of disaster; mercury poisoning in Minimata, Japan, the Union Carbide fire in Bhopal, India, and the Exxon Valdez disaster in Alaska are well-known examples which caused severe local problems. Hazardous materials and wastes are on the increase across the globe, with estimates varying from 375 to 500 million tonnes produced annually (Middleton *et al.* 1993), and many believe it inevitable that further problems will occur as this increasing volume of hazardous waste must be transported and stored for disposal.

Hopwood (2001) notes that while dirty obvious pollution has declined in the developed world, the more insidious forms of pollution which affect the food chain and water cycle remain in both the developed and developing world; 20% of the world's population lacks access to safe water. Chemical entrants into the food chain have been well-documented and their effects on both nature and humanity have been studied. Reduction in sperm counts, increases in exposure to hormones and damage to immune systems have all been linked to industrial and agricultural pollutants, much of it long term and slow in build-up, causing possible cross-generational problems as well as the more obvious immediate poisoning cases which are more publicised. All of this is long term and difficult to measure, predict or track.

Pearce *et al.* (1990) outline the difficulties in measuring and policy-making on pollution. They note that many pollutants are not polluting until a threshold has been achieved and setting levels of safe use is notoriously difficult. On the one hand, gradually increased effects are often difficult to measure, but, on the other, 'better safe than sorry' solutions based on the worst case scenarios can be prohibitively expensive.

Export of waste or of polluting forms of industry from developed countries to developing countries is a global industry although it appears to be in the decline. A poor environment leads to poor health and Middleton *et al.* (1993) point to the huge disparities on access to health services between developed and developing countries, leaving the importing countries with health problems which might not occur in the country of origin. Increasingly health is being viewed as an important indicator of humanity's interaction and degradation of the environment. Pearce *et al.* (1990) note that richer countries are often involved in importing sustainability through this type of activity, i.e., improving their environment and improving the quality of life by removing waste from their own locality, transferring polluting industry elsewhere or taking resources from abroad, while the overall effect globally is a net loss of sustainability.

Loss of natural resources

The possibility of limits to use of natural resources has already been noted at the start of the chapter, and the consequent loss of resources such as mangroves and coral reefs has been well-documented and visible. Coral reefs, in particular, are a high-profile loss mined for materials and polluted from a number of sources. Mangrove swamps are an important resource for humanity, as breeding grounds for 90% of the world's fish harvest by weight. However, it is estimated that 56% of the world's mangroves have been lost since pre-agricultural times, through man-made changes of river flows, overuse, pollution and trans-formation to agricultural or urban land (Middleton *et al.* 1993).

Other basic materials are likewise destroyed or consumed in ever greater quantities. Weizacker *et al.* (1998) point to the absolute waste in the production of even basic commodities suggesting that efficiency alone (rather than just scarcity) should be a strong driver for less use (Table 3.2). An interesting angle on basic commodities is the issue of control, important to the advocates of sustainable development who see transparency and democratic control as part of the equation. It has been reported that the production or extraction of many basic commodities are often monopolised industries, typically with groups of less than six large corporations (Middleton *et al.* 1993). Examples have ranged from agricultural products such as wheat (85–90% controlled) through to ores such as iron ore, copper and bauxite (80–85% controlled).

Table 3.2   Waste in the basic material supply chain (after Weizacker *et al.* 1998).

Commodity	Material moved or processed to produce 1 kg of commodity (tonnes)
Sand and gravel	0.65
Cement	10
Iron	14
Phosphate	34
Gold	350 000

It is, however, the question of exhaustion that taxes many environmentalists, although Lomberg (2001), again, cautions about the belief that limits are real or immediate. He quotes the known oil reserves which now stretch to 150 years at current consumption, although this itself suggests that there is a limit. Likewise, known reserves of cement, aluminium, iron, copper, gold, nitrogen and zinc have all grown by factors of between 2 and 10 in the past 50 years. Together these account for 75% of global expenditure on raw materials.

In conclusion, therefore, there are elements of the environment that suffer from human activity, much of it insidious and difficult to measure. There is much debate, however, on the nature and full extent of the damage although all players appear to accept that substantial damage has occurred. Climate change is real, pollution is an accepted evil while the loss of biodiversity and natural resources present specific problems.

3.1.2   The extent of damage is still being debated

While the previous section highlighted the headline environmental damage which most parties agree affects the globe, there are other areas where there is a more active debate on whether there is a problem or not. This is often the result of a lack of visible or quantifiable evidence, or it may be that political consideration causes the problem rather than environmental limit or needs. This sometimes occurs where the problem is localised to a remote location, where collection of evidence is difficult or where the cause of the problem and the effect are in different locations, making connecting arguments hard to quantify.

Overuse of water/groundwater

Water is an abundant resource on the planet. Its geographical distribution can, however, cause problems. Overuse of water/ groundwater is an increasingly important issue for many nations, leading to tensions and the threat of war over water rights. A significant fact in this debate is that 15% of all nations receive more than 50% of their water from areas in neighbouring nations. The damming of rivers upstream preventing or reducing supplies downstream is a major source of problems in areas such as the Middle East where Israel, which controls the River Jordan, is often at the centre of disputes (Mylius 2000). A 1992 report suggested that 32 countries were in some sort of dispute over access to water rights (Middleton *et al.* 1993).

Distribution of water and rainfall remains, however, the most significant concern. Although there is sufficient fresh water globally to meet all current demand, it is estimated that over 60% of humanity will live in a water-stressed area by 2025 (defined as consumption being more than 10% of renewable supply). It is further believed that 80% of all diseases and one-third of all deaths result from contaminated water (Middleton *et al.* 1993).

Salinisation and desertification

The overuse or misuse of soil leads to a number of problems, such as salinisation and desertification. The United Nations

Environmental Programme (1999) estimates that 1 billion ha of dryland has suffered from degradation to varying levels. Goudie (2000) suggests that countries such as Australia, China and the USA have 15% or more of their irrigated land affected by salinisation. This affects productivity of the land, possibly bio-diversity and the ability of the local population to sustain their economy.

Consumption and waste production

The effects of consumer societies, with high levels of consumption and high levels of waste, have been well documented by many environmentalists. Ecofootprints are a tool devised by Wackernagel and Rees (1996), who have attempted to define the land area required for the supply of a community with all its needs and to absorb all of its wastes. It is a very simple concept, very powerful visually, but limited by information availability and the consequent accuracy. This causes problems when comparing rather than developing rough estimated absolutes.

Much of the work associated with ecofootprints has concentrated on cities since they are seen to be major causes of sucking in resources from elsewhere, urban sprawl and sources of huge amounts of waste. Girardet (1999) has used the concept to assess London. Its inputs include 20 million tonnes of fuel, 40 million tonnes of oxygen, 1 billion tonnes of water, together with millions of tonnes of food, timber, paper, plastics and many other materials. Its wastes include 60 million tonnes of carbon dioxide, 7.5 million tonnes of digested sewage sludge and 15.3 million tonnes of industrial and civic wastes. Although London's surface area is 158,000 ha the combined effect of the above is to create a footprint for London of 19.7 million ha (125 times its surface area), an area that is close to the size of Britain's surface area.

Lomberg (2001), concentrating on the 'waste' end of this type of argument, suggests, however, that individual perception of this problem needs to be placed in context. The US production of waste (generally taken to be the world's most wasteful society) over the next century will, even with the worst estimates, cover an area $28 \, km^2$ or 1/12,000 of the area of the entire USA, still a substantial volume. Hopwood (2001), using World Bank figures, has shown how the US energy use for a population of 275 million equals the total energy use of the eight most populated third-world countries with a combined population of over 3 billion people (Table 3.3).

Table 3.3 Energy use footprints (after Hopwood 2001).

Country	Population (millions)	Energy use per capita (kg oil equivalent)	Footprint (energy × population)
USA	275	7937	2 182 675
China	1242	830	1 030 860
India	980	486	476 280
Brazil	166	1055	175 130
Mexico	95	1552	147 440
Indonesia	204	604	123 216
Nigeria	121	716	86 636
Pakistan	132	414	58 080
Bangladesh	126	159	20 034
Total less USA	3066		2 117 676

Again, this points towards wasteful use of a commodity, which is inefficient and polluting to the earth.

3.1.3 Environmental damage may be reversible

Having looked at issues of environmental damage where it is difficult to reverse the problem or to develop a consensus to prioritise the problem, this third section looks at environmental damage where the evidence appears to show that damage can be reversed or the situation can be improved. This is often because the scale of the problem has not reached a threshold beyond which permanent damage has occurred. In other cases, typically in the developed nations, governments and the private sector have committed resources and effort in the past to prove that even apparently intractable problems can be reversed.

A recent book by Lomberg (2001) studies the area of environmental improvement. He suggests that air and water quality have improved in many areas. Examples of this include London's air in the 1990s being the cleanest for 400 years, lead emissions in the UK reduced by 90% between 1980 and 1995 and sulphur dioxide levels reduced by 60% in Europe in the same period. Middleton *et al.* (1993) also provide figures showing how urban air pollution has significantly decreased in high-income countries, although they have remained relatively static in middle-income and low-income countries, indicating that the application of money and political will can make a difference (Table 3.4).

Lomberg (2001) further suggests that waste production in Europe is becoming less toxic, with toxic chemicals in the North

Table 3.4 Air pollution across developed and developing nations (μg of particulate matter per m^3 of air) (after Middleton *et al.* 1993).

	1979–1986	1987–1990
Low-income countries	323	337
Middle-income countries	160	152
High-income countries	625	63

Sea reduced by 76% and ammonia in rivers down 50% in the last decades of the twentieth century. Worldwide, oceans are cleaner, with oil spills greatly reduced in the past two decades, forest cover has increased between 1950 and 1994 and biodiversity is even showing signs of improvement with whales, bald eagles and elephants all off the endangered species list.

Much of the Lomberg arguments are controversial, with his evidence open to accusations of being selective or the science not being exact. However, many of his opponents do the same and, importantly, he provides a useful reminder that some of what has gone wrong can be rectified. How much can be rectified, however, is a difficult question.

Middleton *et al.* (1993) point out that the technology exists to reduce energy consumption in the developed world from its current average 7 kW per capita to 3 kW per capita (in the developing world the average is 1.1 kW per capita) and to deal with the inefficiencies of power stations. However, energy conservation would lead to reduced sales and producers fearful of the negative effect on their competitive position and financial standing in the market. Thus, although technology exists there are strong financial disincentives to make the improvements.

At a local level, environmental damage, which may be rectifiable long term, can be very severe short term. An important case at present is the ecological disaster surrounding the Caspian Sea, most of it attributable to human activity (Brewis 2001). Pollution from oil activities, industrial waste, nuclear waste and farm pesticides and over-farming of land and the sea have all contributed to a loss of biodiversity. There has also been damage to human health, with incidences of stillbirths, blood diseases and tuberculosis all dangerously high.

Examples like this are often the result of local mismanagement, although wider trading and economic patterns play their part.

Another area of primary concern is the effect of local population pressure. In the last part of the twentieth century the megacities of

Asia have been highlighted as a potential environmental cata-strophe in the making (Jacob 2001). Air pollution in the larger cities in the region was amongst the highest in the world and caused 100,000 premature deaths in India, Pakistan and Bangladesh alone. The region will become the largest source of greenhouse gases by 2015, with 50% of the region's population living in cities (Jacob 2001).

Thus, the quantity and concentration of human beings in one area of the globe may therefore be a further factor to consider. This factor is not restricted, however, to poorer areas where the problem is most visible but extends to more developed nations, although, as noted earlier, the patterns of consumption and waste production can be better hidden.

3.1.4 Comment and analysis

It was noted in Chapter 2 and again at the start of this section that the nature of the evidence, its source and its robustness are as important as the conclusions that can be drawn from it. Both economists and the environmental lobby have problems with their evidence. Table 3.5 compares the problems of economists (noted in Chapter 2) with those of the environmental lobby.

Table 3.5 Problems of evidence.

	Economist	Environmental
Evidence type	Often indirect measure	Often direct measure
Definition of issue	Selective, optimistic?	Selective, pessimistic?
Quality	Narrow?	Complex?
Quantity	Narrow?	Complex?
Major issues	Dealing with evidence thresholds, time-lags, extrapolating current trends, costing impacts	

The evidence available to environmentalists, despite its scientific appearance and despite agreement that damage is occurring, suffers from the effects of complex natural systems and the difficulty of gathering evidence on such dynamic systems. Many of the effects, for example, are hidden within the water cycle or in the food chain, which is notoriously difficult to fully map. Many forms of pollution have limited effects until a certain point when saturation occurs and widespread damage results. Predicting saturation point can be very difficult, but it is often important to act before this occurs to avoid catastrophe.

There are many questions surrounding the costing of damage versus benefits. An example is 'El Nino', a naturally occurring climate pattern which seems to be changing in frequency and intensity, causing weather problems. Lomberg (2001) quotes a study of El Nino in 1997–1998, which temporarily caused temperature changes in the Pacific Ocean and was credited with increasing storm frequencies, warming temperatures and causing irregular weather across the USA. Estimates of the damage were put at $4 billion and were widely publicised in the media. Less publicised were the benefits of warmer winters and less severe weather elsewhere, estimated at $19 billion.

The evidence coming out of both camps suffers from common problems despite it being of different form. This suggests that evidence needs testing, and priorities need to be properly explored. The 'export' of problems obscures many issues for both camps, making it difficult to apportion blame or to correctly apportion cost. Many issues such as overuse of water and soil are complex subjects where much of the blame may appear to lie at local level, although wider trading patterns may also play a part. Waste, on the other hand, is a universal problem. This leads to a situation where the priorities are muddled and factors such as overuse of water, rather than becoming an overriding priority for humanity, tend to be reduced to secondary importance.

The development of choice is not helped by selective use of data and the assumptions that support the analysis of such statistics. This is a problem common to both sides of the argument, with overestimates of damage frequently attributed to the environmentalists and underestimates attributed to economists who are against change. Pearce *et al.* (1990) point to preference and priorities which differ across the world as a major force resisting any consideration of radical approaches.

The failure to clarify definitively and to quantify explicitly leads to mixed up priorities and an inability to properly agree the cost of change or no change.

3.2 Political and scientific perspectives

As stated earlier, there are a range of views that could be described as environmentally led. Hopwood *et al.* (2000) have produced a useful mapping of the cast of players in the sustainable development debate (Fig. 3.1). There are a number of ways of dividing the groups in the map and Hopwood *et al.* propose dividing it into three camps: status quo or those who associate most closely with the traditional systems seeking little change, reformists who see the need for reform of systems to deal more fairly with social issues, conservation or the environment, and transformationists who see mounting problems which require a radical rethink of the world order. The mapping will be used to aid discussion in this and the next chapter. This section will concentrate on those schools of thought that lie close to the environmental axis (the bottom third of the graph) or on the outer edges. In the next chapter some of the middle ground will be further analysed.

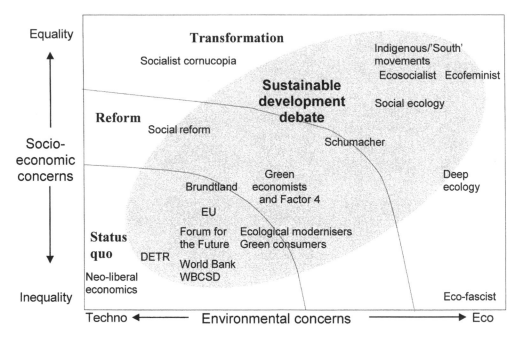

Fig. 3.1   Mapping sustainable development views (Hopwood *et al.* 2001).

The green end of both reformist and transformationist schools of thought questions the fundamentals of the belief that growth in economic terms can be sustainable, and points out that increasing use of resources and production of waste will hit natural limits and is fundamentally unsustainable (Hopwood *et al.* 2000). Beyond this one unifying argument opinions can diverge, with Dobson (1995) suggesting a split between conservation (status quo), reform environmentalism and radical ecology, while O'Riordan (1989) suggests a range between those who believe that technology will provide solutions through to those who believe that only nature will win in the end (Table 3.6).

Table 3.6 Comparing environmentalism and ecologism (after O'Riordan 1989).

	Environmentalism (human-centred?)	Ecologism (ecocentric)
Philosophy	Humanity controls and masters the world to make it more certain	The earth provides the system and should be treated with respect
Label	Dry/shallow green	Deep green
Policy	In favour of current economic and political systems with some revision. Self-regulation priority	Fundamental changes to economy and political structure needed. Redistribution priority
Management approach	Reliance on science and experts, markets and economic logic	Based on biodiversity, new conservation strategies and environmental science

All schools of thought contain a mix of science and political belief. Many of the green lobby, for example, would see the loss of connection between humanity and the natural world, most obvious in urban living, as being a core source of problems, with proposals to correct this ranging from a slight altering of the status quo with the greening of cities through promotion of 'human-scale' communities and into the more radical end advocating a return to subsistence-style living (Goldsmith 1992). Sullivan (2001) notes that the policies of even some of the USA mainstream environmental groups would lead to wholesale removal of rural communities from the middle of the USA, destroying lives and livelihoods to create a nature wilderness for urban dwellers to enjoy as a leisure activity.

The earth, as an ecosystem independent of humankind, features heavily in this literature (Lovelock 1988). Much of it is science-based, with evidence coming from systems-based approaches (see the case study on climate change in Chapter 7). These complex models map biological and non-biological systems with feedback and balances in which the earth strives to achieve stability, a real but difficult objective.

Green transformationists often argue that humans, social and economic concerns are immaterial to the future, and only ecology is the one true focus. Dobson (1995) considers the fundamental concerns in ecology to be:

- There are natural limits to human action which cannot be superseded by technological change and these need to be accepted.
- Morality is based on the natural world rather than any modified stance from science.
- An ecocentric approach is a principle, with the earth as the centre of existence rather than humankind.
- There must be a tendency towards an assumption of damage in humanity's action (possibly equivalent to the preventative principle).
- There is an assignment of some sort of rights to life, species and ecosystems.

Others would argue that non-human species should have similar basic rights to humans, and that natural systems and biodiversity have an immeasurable value (Naess 1989). Naess suggests that the flourishing of human life might benefit from a reduction in human population, but the flourishing of non-human life is dependent on the decrease. Further, diversity has an intrinsic value which must be recognised. He suggests an obligation on those experiencing nature to realise and participate in improving the current situation.

The subjects of land ethics and attitudes to wilderness raise many interesting issues. The more extreme would advocate active population reduction and see natural disaster as a useful support of sustainable development. Moreover, they call for a complete stop to further development. Supporters at this end of the radical, deep-ecology movement verge on an anti-human theme, and are often closely associated with direct action groups (Hopwood *et al.* 2001). This level of ecological intensity is often labelled eco-fascism, since although the professed basis of morality is eco-centred it is an eco-centred approach based on interpretation by a narrow group of humans. There is also a tendency to ignore environmental issues that do not fit a romantic view of the wilderness (Mellor 1997).

Leopold (1966), by contrast, argued that what is needed is a system of conservation based on the interaction between land and its human inhabitants. Budiansky (1995) notes, however, that the idea of wilderness and the protection or restoration of landscapes untouched where man has only visiting rights has still led, in some instances, in the USA to residents of areas declared as wildernesses being expelled from their homes in order to develop the wilderness (Spence 1999, Wood 1995).

The *reformist school* views, by contrast, are centred more on a

human viewpoint, with environmental issues one aspect of the
bigger picture. There is a greater acknowledgement that humanity
is a part of the grand scheme. As such, some parts of the current
approach to development can be modified for improvement, other
parts need the sort of radical change that transformationists would
support.

They are likely to share with transformationists the belief that
the ecosystem imposes limits to growth both in terms of resource
depletion and the earth's ability to cope with waste and pollution.
There is a strong science theme with many of the views suggesting
that individual environmental effects can be split out, assessed,
costed and adjustments made to current systems to make allowance or curb specific action causing the problem, linking them to
the reform end of economics (Pearce *et al.* 1990, Hawken *et al.* 1999,
Daly and Cobb 1989). Some of this has already been studied in
Chapter 2 (those who saw the driver as markets) and some are
studied in more detail in Chapter 4 (who see other drivers of
change).

3.3   Critique and analysis

It is important to note the views of economists of this end of the
debate. In general (*Economist* 2000b, 2000c), the economist wing of
the debate appears to confuse many very legitimate green activists
with the assorted activists who target world trade or development
bodies and multinational companies. Henderson (*Economist* 2000b)
suggests that many see common ground in four main areas:

(1)   Rather than leaving the market to decide, they believe that
      some industries are essential and others non-essential.
      Activists would prioritise the 'essential'.
(2)   Cross-border activity even between private companies
      inherently involves national governments and thus there is a
      public interest in this activity.
(3)   Exports are a gain, imports a loss and tariffs add to employ-
      ment in a simplified equation of what is good for a nation.
(4)   Profit is a questionable concept.

Economists note that many of the messages brought forward by
the diverse groups of protestors to world trade bodies are not
intellectually coherent and, in fact, many are mutually incompa-
tible. Many are based on false information (an often quoted case is

the Greenpeace attack on the Brent Spar which badly exaggerated the level of pollution on a disused oil platform) and many promote transparency despite being themselves opaque and unaccountable.

One consistent target has been the oil industry, seen as a source of much of pollution, lack of transparent decision-making and the main contributor of products that create carbon dioxide. Many of the target companies have, however, started to look at their operations and promote alternative products or greener ways of working despite this scepticism (Dunne 2001).

The media perspective

Both sides hark back to class wars, with the environmental issues caught in the middle, reduced to secondary importance. Appleyard (2001) refers to a belief that the oil shock of 1973 led to a deliberate policy from the world's elite. The leading developed-world governments and multinationals combined to promote a policy of mass unemployment and shift the business emphasis to money-making from market manipulation rather than production of goods.

The rise of the 'brand' and the globalised corporation was identified by Naomi Klein as a significant marker in this change. These corporations supported by world trade bodies have overtaken national governments in becoming the powerful drivers behind the global capitalist system, inflicting unnecessary social change and environmental damage. The solutions proposed are numerous but include ending the limited liability company, opening up business to being accountable for whatever damage it is perceived to have done (Appleyard 2001).

All challenge the basic free market assumption of capitalism and Henderson (*Economist* 2000b) actually believes that national governments are natural supporters of these beliefs. It is only in periods when these beliefs are shown to fail (such as the early 1990s and the collapse of communism) that liberals are in the ascendancy. Henderson believes that this has been the case throughout modern history and the only new aspect has been the growth of NGOs and the marketing of antiliberalism as humane, rights based and environmentally friendly.

It is further noted that the momentum behind green politics is such that it is important that the general thrust of the argument is right or else problems may quickly arise. The arguments behind green politics appear careful and politically correct. Green political parties have had success across the developed world, most notably in France and Germany where they have formed part of the government.

Lomberg (2001) highlights three factors that need to be carefully considered in reviewing green arguments: (1) scientific funding is biased towards problems and it is therefore beneficial to overstate the problem; (2) environmental groups need the media to sustain their momentum and the media is selective in its reporting (disasters are good, good news is bad); and (3) people's preconceptions are often, but not always, wrong and many preconceptions involve green politics as selfless and altruistic.

Economists suggest that calling a halt to the use of resources that are not in danger of depletion may be wasteful. However, they also note that many of the arguments are based on a little bit of truth and are couched in very pleasing terms – human, environmental, green, participative – so that they cannot be easily dismissed. Their effect on 'normal working' of multinationals and trade bodies is effective and noticeable. The promotion of 'fair trade' coffee, the backlash against sweat-shop labour for brand names, protests at the world trade meetings and reversal of decisions on economic development or infrastructure provision have all progressed the cause but may not always help the progress of the nation involved.

3.4 Summary

There is little doubt that environmentalists have rightly raised the issue of environmental damage as a concern. The degree of damage and the priority that it should be accorded remain areas of contention. There are serious issues involving the amount of evidence, problems in identifying key evidence thresholds and persuading other schools of thought of the importance of this evidence.

On the issue of limits to growth it is clear that, very simply, there is a finite limit to the use and misuse of natural resources and therefore the arguments at the heart of environmentalism in whatever its form are correct, whatever the time-scale. At issue, however, are the scale, priorities and correct analysis of cause and effect in each of various environmental issues, as suggested in the first section of this chapter. At the same time, the issues of equity and control, while secondary to the fundamental science of the subject, nevertheless add to the concern.

The solutions put forward by the various shades of green suggest varying degrees of punishment for those held to be responsible by environmentalists. However, evidence is crucial if corrective action and responsibilities are to be clearly identified. The collection of perfect evidence demanded by many is difficult to achieve in practice and so there may be a need to consider the threshold of evidence required to change policy, stop damage or take rectifying action. Ecological solutions that involve fundamental change will involve fundamental mistakes if they are wrong, but, equally, may provide the brightest, quickest solutions to global problems if they have been correctly assessed.

If there are serious environmental concerns does it follow that sustainable development is at risk? Or can environmental issues be broken out and dealt with separately? Some of the evidence suggests that it can, although much of it suggests that the mainstream economist approach of needing a sufficient weight of evidence and then dealing with the environment as an 'equal' priority sector leaves room for improvement. The evidence that environmental problems are at a catastrophic stage is not overwhelming, but it is, nevertheless, worrying, with many of the arguments pointing to two key issues raised in Chapters 1 and 2: the lack of a future perspective (futurity) and displacement, or passing problems elsewhere.

More importantly, would the separation of environmental issues agree with the spirit and intention of the sustainable development debate, or is it simply displacing the problem (placing it in a corner and assuming that others will deal with it)? All of these are issues that will be addressed again in Chapters 5–8.

4

Some of the Shades In-Between

It was noted earlier that the original polarity in the sustainable development debate may have been between mainstream economics and the edges of environmentalism or ecology (Pearce *et al.* 1990). The roots of the discussion from these two schools have thus been covered in the previous two chapters, an appropriate starting point since these ends of the envelope of beliefs have generated most debate. However, the subject is complex and has branched out in a number of directions, beyond pure support of traditional mainstream economics on the one hand and environmentalism on the other. Beyond the very early formative years of debate it is, in fact, unlikely that such a simple choice of economics v. ecology was ever possible.

There are a number of schools of thought that have developed between the two ends of the envelope, often based on the one specific strand or shade of reality. The array of theories and beliefs available very quickly confirms that sustainable development remains a contested concept (Hopwood *et al.* 2000). A sample of three of these areas of debate is therefore studied in this chapter.

(1) The *equality–inequality* theories: the response of the social sciences to the early debate. Some of this has been touched upon in Chapter 2 since economics has an influence on this. Within Fig. 3.1, this area of debate forms an area close to the y-axis. Much of this debate concentrates on the effects and trade-offs that economic growth has for the development of society in general.

(2) The *techno-centred* theories: the response of engineering sciences and technology to the early debate. Again this cuts across many of the issues raised in Chapters 2 and 3. Many technologists would probably sit alongside the traditional economists towards the bottom left-hand corner of Fig. 3.1, with the belief that technology will, with some help, improve development.

(3) The theories of *balance* which try to attain a model with no obvious bias in either social, economic or environmental directions. This would follow the line of Brundtland, the

World Bank or the DETR in the UK, although many of these may actually sit closer to traditional economics than a true balance.

There is a fourth area of growing interest, the belief that theories cannot be universal and must be specific to a particular situation. On a global scale, an obvious example of this is the result of differences between views and priorities in developed countries and those in less-developed nations. This will be examined in this next chapter by looking at the lessons from practice, the source of much localised theory.

4.1   Basic assumptions and evidence for it

Unlike the previous two chapters, there is no central argument around which views are formed and then diversify. Different areas of debate are grouped into three in this chapter, but there are no clear or distinct boundaries between them. This is particularly true of the theories in the section on the middle ground, since balance is a difficult objective.

4.1.1   The debates on equality–inequality

The fundamental belief across this area of debate is that humanity is the centre of the debate. The debate probably includes positions across transformation, reform and status quo, and an important theme is the importance of political stance in the solutions proposed.

An interesting approach to the explanation of political stance comes from the political compass (Fig. 4.1), where it is suggested that left and right are essentially economic stances on market choice while the social dimension ranges from state control through to freedom. Elements of the four directions are encompassed in any political choice. Mapping an individual or state onto a point on this graph determines its initial state. Advocates of the status quo are thus likely to want to remain at the same position, reformers will want to move slightly, while transformationists will probably see the need to move to another quadrant of the graph.

Many advocates of the *status quo* believe that changes to society that are already in train, such as better information management, more efficient technology, more effective government policy or

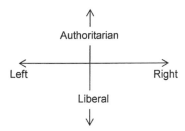

Fig. 4.1   The Political Compass (After One World Action (2001)).

tighter scientific knowledge of the natural world, will combine to achieve sustainable development.

More informed consumer power will keep business on the path to an improving ethical stance and push government in the direction of delivering policy for righting gaps where market forces are providing the wrong balance. All of this will marry the needs of social, economic and environmental concerns, although the environment is afforded no special treatment. In many of these circles sustainable development is interchangeable with the concept of quality of life. Government policy, where required, is advocated more in terms of environmental protectionism, beneficial tax credits for positive action which assists sustainable development and commitments to ethical behaviour. Assisting people to help themselves is often a theme, although this can range from a very active focus from socialist-leaning groups to a more passive focus from others.

Much of the debate is primarily concerned with the human side of sustainable development, with one view being that happy thoughtful humans look after their environment. An important issue in this is the equality of opportunity, believing that the current system, though imperfect, represents the best form of equal opportunity for most people.

The *reformers* see the current system as fundamentally flawed and promoting more rather than less inequality (Table 4.1). They see the manifestation of problems arising from inequality, and many would argue that poverty and most environmental problems arise from poorly managed capitalism. A critical aspect of this for many is income inequality which is seen as detrimental to social cohesion and ultimately to humanity and its environment. In the USA the average income of the wealthiest 1% increased by 142% in the period 1979–1997. In the same period the average income of the bottom 20% reduced by 3% (Sullivan 2001). Much of the blame has

Table 4.1   Income inequality in 1999 (Source: World Bank 2001).

	% world population	% gross income
Low-income countries	40	3.3
High-income countries	15	79

been pointed at globalisation, and low-skill jobs being transferred quickly across borders, away from richer nations such as the USA and into lower-income nations.

The simplest mechanism for dealing with income inequality is wealth redistribution through taxes. It is reported, however, that this has not improved the situation in the USA (Sullivan 2001). The tax burden of the richest 20% of the US population increased from 57 to 65% during the same period, 1979–1997, while the burden of the lowest 20% decreased from 2 to 1%. Thus, there are no easy options. Over the same period the USA created more wealth and more employment than almost all of the rest of the world combined, much of it in high-skills work, attracting inward migration from across the globe, and creaming off the well-skilled from other nations.

At the global scale the share of the world income going to the poorest 10% of the world's population fell by over 25% between 1988 and 1993 while the richest 10% saw an increase of 8%. Technological change and financial liberalisation associated with globalisation appear to be the main factors behind the growing disparity. Again, a link is suggested between poor average incomes leading to a lack of resources for proper governance which cause problems with environmental issues (Wade 2001).

Middleton *et al.* (1993) highlight evidence of the USA and EU destroying jobs or markets elsewhere because of their agricultural policies. Between 1980 and 1984 the EU and USA raised their share of global sugar trade from 17 to 28% through use of subsidies, to the detriment of unsubsidised and therefore presumably more efficient sugar cane grown in developing countries. In 1986 wheat grown with a $100/tonne subsidy was sold to countries such as Mali for $60/tonne, destroying what was assumed to be more efficient competition. In all of these cases it is easy to blame the rich for competing unfairly against the poor, but it is clear that systematic failures and muddled priorities are as much of a problem as selfish single nation or corporate behaviour.

As with the debates within environmentalism discussed in the last chapter, many of the views covered within the social–

environmental debates often have politics as a key driver. This applies at both the practical level and a more philosophical level using the thoughts of Marx and Engels to link exploitation of people and environment to capitalism's shortcomings.

Many see a link between left-leaning politics and green politics and thus there is a familiar sight of red–green coalitions in European politics. Lifestyle politics is often seen as an important element – taking the issue down to the level of an individual's choice. At its extreme, it calls for direct action on specific issues or opt-out from current systems by individuals.

The role of the former Soviet Union is an interesting case study for debate. Viewed as a socialist utopia, it should, in theory, have had strong environmental credentials. In practice, it was the opposite with some of the worst environmental damage in the world. Various explanations for this 'anomaly' have been put forward, including simple but basic mistakes in the state's structure, that it was not far enough down the learning curve to accept the importance of the environment, or that there was a lack of conscious informed decision-making and truly socialist credentials (Hopwood *et al.* 2001).

Non-governmental organisations (NGOs) have become an important element of the sustainable development debate and it is believed that there are at least 30,000 international NGOs in existence. Despite their original green credentials many see themselves as the first step towards broader international citizen groups (*Economist* 1999). Hopwood *et al.* (2001) would classify many mainstream 'environmental' NGOs in this category rather than with the deep greens, with mainstream NGOs positioned somewhere in the middle of the debate between mainstream economics and ecology, sometimes working closely with governments and development banks, sometimes closer to the deep greens.

Greenpeace

Greenpeace is perhaps an example of this type of NGO. Originally a body that probably sat close to the deep greens claiming 'devotion to nature above materialism or greed', it has now been described as a 'corporation' run by an ex-government adviser replete with boardroom battles and 'quietly relieved to see the back of the unpredictable firebrands who gave it its crusading edge'. Others, however, see it as likely to survive only by becoming more extreme, targeting ever more conventional goods and services, some of which may do more good than harm (Driscoll 2001).

What do they see as solutions? In the past the big issues were simple and environmental in nature: nuclear testing, dumping toxic waste and more local issues such as whaling, many of which are now seen as unacceptable activity by the majority of people. The more conventional now point to changes in priority to include a greener agenda, proactive approaches to increasing the trickle-down effect that is said to result from today's system, and policy reform at national level to focus on human well-being rather than abstract pure economic terms. Some of the NGOs are involved in actual delivery of policy or services designed to alleviate the worst excesses of whatever stream of non-sustainable development they have targeted.

Many, however, continue to see themselves as apolitical. Technical groups have become an important part of the NGO scene, often co-opted into making policy because of their knowledge of detail in their given field (*Economist* 1999). Their ability to influence and communicate appears to have greatly improved with the development of new modes of ICT.

However, whatever the political or non-political nature of the core theme, the democratic accountability of these organisations is increasingly being questioned. They have been viewed as self-appointed, vocal and one-directional, and have been accused of squeezing out the less mobile and less affluent, the very sets of people that many would hope to help (Driscoll 2001).

4.1.2   Some other socio-environmental theories

While there are a few environmentalists or ecologists who focus strongly on the ecology/environment aspects of the debate almost to the exclusion of the socio-economic, there are many who see social issues as equally important in the debate. The simplest to map advocate a return to subsistence. Ideally this is seen as a return to nature and community avoiding the environmental damage and consumerism of current systems and, at the same time, promoting the re-emergence of simpler community values. Others would point to problems of social justice, social ownership or equity in its many forms (race, nation, gender) as being paramount in a better system. Many would see a link between a mounting environmental crisis with a continuing social one, with both being the visible effects of a similar cause, a lack of equity in the current systems. Middleton *et al.* (1993) report for example that 80% of sub-Saharan food is produced by women but yet they are denied basic equal rights. However, with environmental schemes in the same areas, often 50–90% of the volunteers are women, mindful of the importance of the environment to their livelihoods.

One belief is that people having control over their lives, resources and environment reduces inequality and environmental degradation. Many in this circle suggest that democracy is a crucial factor, and would query the accountability and transparency of many of the development banks, multinational companies and the range of national and local government institutions, from the old Soviet Union through to the western democracies. One corner of this, for example, is social ecology, which has many branches, but the core sees humanity and nature rooted together (Pepper 1993). Ecosocialists see the exploitation of people and environment through global capitalism as the main cause of problems, and the only solution as the abandonment of capitalism and its economic structures. Ecofeminists propose a link between the subordination of women (who have a closer affinity to nature than men) and degradation of the environment (Mies and Shiva 1993). All see the breakdown of the link between humanity and nature through production techniques and life-styles which have become detached from nature.

Many would note that there are differences in outlook between northern (developed) nations and southern (developing) perspectives. While campaigns from this school of thought in the north tend to reflect idealistic approaches to improving life for all, the harder focused campaigns in the south revolve around the rights of grassroots bodies to live life to the full rather than struggle with the results of environmental, social and economic failure. Hopwood *et al.* (2000), for example, suggest that sustainable development proponents include Brazilian rubber tappers, because of their struggle for rights, the Ogoni people of Nigeria through their differences with western oil companies and corrupt government, and the Zapista uprising in Mexico. All these movements started in a battle to improve environment and social justice. It is this axis that has the strongest link with mainstream political thought – the arguments of capitalism versus socialism or authoritarian versus libertarian, which often leads to the typecasting and all the associated baggage which is associated with politics.

Beside the relatively straightforward debate on equality–inequality a twist on the debate has been the argument about the disappearance of social capital. Many bodies such as the World Bank, the EU and national governments now encourage a belief in the concepts of social and environmental capital to sit alongside the more familiar economic capital. The stock of each is important to the survival of the Earth and humanity. Economic capital, the most well-known of these, has its obvious difficulties when measured through GDP as explained in Chapter 2. Environmental capital measures, again discussed in Chapters 2 and 3, are in their infancy.

However, social capital may represent the most difficult of the three to conceptualise and measure. While both economic and environmental capital have an element of physical, measurable stock, the concept of social capital, by contrast, has a basis in less visible properties. Puttnam (*Economist* 2001a), as noted earlier, has suggested a method of quantification which has received widespread attention. He points to a decline in communal behaviour in the USA as being a strong marker of the loss of social capital in even an economically strong nation. It is interesting to note that the editors of the *Economist* (2001a), in response to Puttnam, suggest that the decline in communal behaviour is a problem resulting from government crowding out civil society. They believe this is the natural propensity of a government promoting a welfare agenda with better income and access equality, presumably one more argument in favour of smaller government and freer markets.

This very brief summary of the equality–inequality debate within sustainable development indicates that political belief is an important factor in determining how to define the issues and priorities. Hopwood *et al.* (2000) provide a useful reading list for further, more detailed study.

4.1.3   The techno-centric arguments

The techno-centred debate is often the domain of scientists and technologists. In contrast to the previous section, where many see choices arising from sustainable development as inherently political, the techno-centred debate seldom appears to acknowledge political choice as being an important contributory factor. The optimists in this school believe in the power of technology and its ability to improve sustainability, pointing, as noted in Chapter 1, to past performance and man's ability to overcome past problems. In particular, the environment is viewed as an area ripe for improvement or protection through technology and through the development of new markets.

The more thoughtful agree that the paths to better incorporation of environmental issues and products need better support, funding and new ways of thinking so that they become mainstream rather than viewed as quirky, interesting but uneconomic. A good example of this school of thought has come out of the 'Factor 4–Factor 10' types of argument. The idea is based on setting an objective of increasing resource productivity by factors of between 4 and 10, and arose out of work at the Wuppertal Institute in Germany and the Rocky Mountain Institute in the US (Hawken *et al.* 1999).

The Factor 4 theories concentrated on resource efficiency but acknowledged the need for new tools for measuring business efficiency, innovations in business practice and some change to public policy. The more stretching Factor 10 theory needed to add in a cultural shift and to address the restoration of natural capital (the improvement of nature's diversity or abundance) for the target to be achieved. While the targets sound ambitious, the authors point to the fact that human productivity improved 200-fold in some industries between 1750 and 1820. They also note that there are natural limits to material availability which will force change and that there are gross inefficiencies in the current systems which can be easily addressed.

The Factor 4–Factor 10 theories are interesting, however, because the difference between the two highlights the difference between the two schools identified in Fig. 3.1 (Hopwood *et al.* 2001) as status quo and reform. Factor 4 promotes a market-led approach adapting and setting new markets to improve the situation but maintaining the status quo. Factor 10 introduces the reform school, where reform beyond markets is more clearly identified. Thus, politics creeps into even the arguments of the technologists. Another factor in both arguments is the acknowledgement that compromise is required in any final solution.

More importantly, they identify a key factor to force the culture change required. The pre-industrial revolution was powered by the need to deal with scarcity of human labour but abundant natural resources. That has now changed to a situation where people are an abundant resource but natural material is becoming scarce, demanding new ways of working. This is an interesting, even if contested, viewpoint.

The gross inefficiencies documented by Hawken and colleagues are very telling in themselves. They record, for example, that 6% of the materials that flow through the US economy actually end up in end-products (implying that 94% is wasted), 99% of the energy put into the fuel tank of a car is lost, 91.5% of the fuel input into a power station is lost before use in an industrial pump, and the production of a semi-conductor produces 100,000 times its weight in waste. Even more striking are the examples of mapping the work that goes into the production of a Cola can so that a consumer can take a small drink and throw the can away.

The move to less wasteful, more natural products needs new assessment tools to both analyse and, just as importantly, allow the public to visualise how new products are better than more wasteful ones measured in more traditional ways. Tools high-lighted earlier such as ISEW and the Ecofootprint add to the picture, but a key tool highlighted in the Factor 4 theories was material input per unit of service (MIPS), which studies all mate-rials bought and sold in the production of a product. To quote from Hawken *et al.* (1999):

'Industry moves, mines, extracts, shovels, burns, wastes, pumps and disposes of 4 million pounds of material in order to provide one average middle-class American's family needs for one year.'

Accompanying the assessment tools is a need to improve business practice, with business initiatives that nudge companies towards ethical, efficient and waste-avoiding systems. Such initiatives link business to responsibilities beyond the bottom line, but which in theory lead to better profits. Such initiatives include Quality Management (which will be discussed further in Chapter 8), Eco-Management and Audit Systems (EMAS), Environmental Impact Assessments (EIA) and a variety of corporate social responsibility schemes.

However, Hawken *et al.* (1999) suggest that such schemes need to stretch to 'Extended Product Responsibility', where customers lease rather than buy and the producer has responsibility for the product throughout its life and is better able to gauge the true costs of the product. A good example of this is the photo-copier which is more often leased than bought in most offices.

Design is seen as a critical stage of the life cycle, with a belief that there is a barrier to the generation of new ideas in traditional teaching techniques. The accepted compromise between time, cost and quality, so often seen as crucial to good project management, is therefore a key constraint to progress (Fig. 4.2). The building industry provides an excellent example. Ninety percent of the

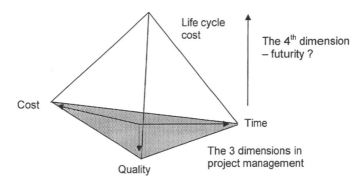

Fig. 4.2 The three-dimensional project management becoming four dimensional.

average American's time is spent in a building. Eighty-three percent of the cost of running a building is employee related. One-third of the energy consumed and two-thirds of the electricity used is through buildings. Thus, buildings are a critical long-term part of human activity and should need careful design. However, after just 1% of up-front cost has been spent, typically 70% of life-cycle costs in a building have been fixed.

This theory suggests that there is a need to rethink the old triangle of time–cost–quality, considered by many good project managers as the framework for considering the main trade-offs required in the design and implementation phase of development projects. Time and cost are economic in nature while quality was often restricted to an initial appearance. This needs a further dimension of futurity possibly through life-cycle cost, although this itself contains elements of time, cost and quality.

Often a key feature in the arguments of this school is the subject of energy use. It is an interesting subject because it has elements of efficiency, use or misuse and the more specific promotion of renewable sources of energy. The energy debate is an important dimension in the technologists' understanding of sustainable development, and they often see the arguments as being non-political since the driver is to fuel the world beyond hydro-carbons, a 'factual' desire, and hence continue socio-economic progress and reduce environmental impact at the same time.

Many of the scientist–technologists fall within the same groups as the more traditional economist view and this has already been discussed in Chapter 2. Most would not query the assumption that growth is good and would choose to work with it rather than against this assumption. Many would classify themselves as conservationists in their approach to sustainable development, with an almost single discipline approach, although each in their own way is multidisciplinary.

4.1.4   The theories of balance

The middle ground suggested in this text arises from the use of the word 'balance' in the definition or description of sustainable development, signifying a desire to pursue the mythical balance between economic, environmental and social needs. This broad definition probably includes the Brundtland Commission, many green consumers and the European Commission (EC). The EC in particular makes use of the word 'balance' in its consultation document for a strategy for sustainable development (Commission of the European Communities 2001). In comparison to the light-green credentials of the UK's Department of Environment, Transport and the Regions and the World Bank, which were discussed in Chapters 1 and 2, the EC has a distinctly greener (and redder) tinge to its policies, a reflection of the political strengths of the greens and the socialists in Western Europe.

The *Economist* (2001b) reports on the European 'obsession' in linking trade and the environment, which is based on the precautionary principle. This is the approach, briefly mentioned in Chapter 2, where action and policy are taken on the basis of preventing damage rather than waiting until it occurs and then acting. Many in the USA, however, suspect that the 'obsession' hides more important protectionism of an uneconomic agricultural sector, although it was noted in Chapter 3 that the USA and the European Union (EU) are both jointly accused of this type of protectionism by other less-developed nations.

Much was made of the EU's part in securing a deal in Kyoto and later in Bonn on climate change, allowing global bodies to work further on tackling the issue (*Economist* 2001c). Further evidence of the green slant to policy is provided through a check of the EU web-site. A search for sustainable development papers reveals that 36% of the documents highlighted are also classified as environmental in nature, while trade, information management, global debate, gender, finance and technology each account for much of the rest.

**The EU
approach**

The Commission of the European Communities (2001) suggests an important set of drivers behind EU activity in the field of sustainable development (Fig. 4.3). These include:

■ A political desire for subsidiarity devolving decision-making down to the lowest practical level so that many initiatives are promoted by groups of municipalities or NGOs across Europe
■ A much stronger green political body within the EU than in many other developed nations
■ Specific issues within the EU such as the 80% urbanised population, an ageing population and a strong political desire to address income inequality.

Fig. 4.3   The means of promoting EU activity in sustainable development.

There are a range of dedicated actions within the EU forming part of a wider strategy for sustainable development. The time-scale for key mile-stones in the development is outlined below:

■ EU-wide economic policy set at Cardiff (1998) – to include economic, employment, social and environmental integration policies
■ Single Annual Review agreed at Lisbon (2000) – bringing together social and economic initiatives
■ Sixth Environmental Action Programme set at Brussels (2001) – to set specific environmental actions
■ Report to Stockholm European Council (2001) – to integrate sustainable development strategy and environmental review

The sustainable development strategy envisaged from this work has six themes based on three priorities and two principles (Fig. 4.4).

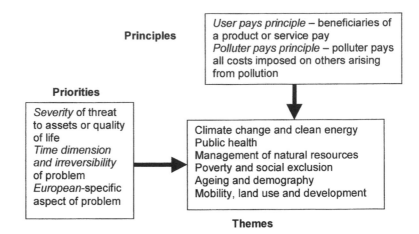

Fig. 4.4   The EU's principles and priorities in sustainable development.

The themes chosen by the EU are very specific when compared to the objectives of other high-level bodies such as the World Bank or national governments. However, they are clearly specific to European issues and remain general enough to cover many issues. It is nevertheless a greener agenda than the USA or the UK, and is matched with a strong research effort devoted to environmental issues.

Devotion to the cause of sustainable development has spawned a huge amount of activity, policy initiatives, expert groups, research and action plans, all providing one facet or another of the EU's approach to sustainable development. Despite the drive towards a single strategy, much of the activity still appears very uncoordinated to the outsider, a reflection of the manner in which the EU takes forward advice and promotes initiatives in all of its work.

Beyond the bureaucrats within the EU there are many groups who subscribe to the idea of balance, as noted earlier. Green consumers, green economists and many NGOs may find a home in this school. Many are idealists seeking an illusionary balance, while other more practically orientated advocates look to indicators as their solution.

4.2   Summary

This chapter has illustrated three disparate groups of beliefs. The politically inclined look at equality–inequality, the scientist– engineers want to study the effects of technology and others are keen to see harmony and balanced solutions. Common to all three is their belief in humanity and its part in the solution. Common too is their belief that by concentrating on one particular part, aspect or driver they can push other pieces of the jigsaw into place, although, equally, however, the economists and the environmentalists share this belief.

With the first group, which revolve around attitudes to social policy primarily, there is a strong link to political choice and decision. If the politicians make the wrong choices in a democracy then another set of politicians takes over when evidence mounts that one or other path to development is incorrect. The linking of social and economic policy is relatively simple in this sense (though it has its grey areas where evidence is clearly a problem), but the connection to environmental policy is less transparent.

Rifkind (2001) has suggested that it is necessary to accept special treatment of social and environmental policy as a premise of policy decision-making because of built-in contradictions between economic modelling and the other two legs. The fact that the techno-centric arguments of the second group generally support the status quo propounded by economists make this a potentially very powerful alliance. Pure science has had both successes and failures in the past, and its ability to bring improvements to the benefit of society and the environment must continue to be fairly challenged if catastrophe is to be avoided. Equally, however, risk-avoidance results in sterile debate and no progress.

A critical issue for both of these groups is the area where the evidence does not clearly support one option in favour of others. Both politics and science have taken humanity into blind alleys in the past.

On the surface, there is less to be wrong in the third camp described in this chapter. A better balance with due consideration of all the elements appears an ideal solution. However, the next chapter examines the reality of a 'balance'. It is certainly true, within these arguments, that separation of cause and effect have, in the past, caused problems. The most obvious example is that of urban living detached from the issues raised in providing food and materials for the city, together with a policy of dumping the waste elsewhere.

Chapters 2 and 3 left the conclusion that the two 'extremes' of economic dominance and environmental dominance leave many questions. This suggested that something in-between may provide better promise. The three areas studied here represent some of that middle-ground, all with very different drivers, and they show the continued difficulty of classification within the sustainable development debate. However, they are important because they address the parts that many of the purists of the first two schools ignore, even if their solutions are poorly developed. Thus economist, environmentalist and those in-between all appear to provide some plausible ways forward, but no one school holds all the answers. A worrying problem remains of 'evidence thresholds'. Pearce *et al.* (1990) suggested the existence of a straight choice between preventative and reactive policies – acting before or after damage has occurred – but it is clear that the choice is not straightforward.

Is it feasible to combine the best parts from the disparate groups of theory? The current system is a working model, but can it be combined with a higher priority for environmental concerns and social safeguards? In fact, EU policy looks remarkably similar to this form, but it is still an imperfect model. Does this equate to sustainable development as a linked, coherent set of ideas or is this not akin to reducing the problem back into three components?

The aim of the next few chapters, therefore, is to (1) analyse practice and what constrains the development of a coherent theory, (2) look at the gaps and possible linkages between important factors in the debate, and (3) dissect this analysis itself to look for further insight by reducing it to the level of the human response to these ideas.

5 Practical Interpretations of the Debate

5.1 The trade-offs in practice

The previous chapters have briefly outlined the many conflicts that result from the adoption of the principles of sustainable development, and the wide variety of theories that ensues. Attention is now turned to lessons that come from the attempts to implement sustainable development in practice. It may be that practice will eventually lead the development of theory, as the lessons of constraints, best practice and achievability provide a framework within which to place the theoretical debate, and this is a theme that is returned to in Chapter 8.

As the first chapter showed, an agreed common framework for sustainable development needs to consider starting points, the process and the end-goals. Practitioners have contributed greatly to the debate, although many of them have tended to spread a story specific to a particular situation or location. This raises issues of the repeatability of the exercise and a query about the universality of the lessons. This can leave an audience unable to see the common threads that purport to be sustainable development.

There are three main areas where the lessons of practice have been very useful:

(1) The identification of trade-offs and the need for compromise, an 'ugly' concept for academia because it lacks theoretical purity although it serves the world of practice
(2) The constraints in practice to what often appear to be easy choices in theory
(3) The theory that does stand up to the challenges of practice, or those that are repeatable across a range of practical case studies.

Much of the illustration for this chapter will be using case study based around the experiences at the Sustainable Cities Research Institute (Mawhinney 2000). A strong theme in guidance on practical sustainable development is to think globally, and act locally (DG-Environment 2000), and the case studies discussed will make strong reference to this theme. The main value of studying such specific projects at a particular research institute is the variation of detail that can be examined. Importantly, the limitations of the Institute's expertise and approaches will be laid out in this initial section so that the reader is aware of the shortfalls and can make his or her own judgement on the validity of the case studies. However, the experience of the author at conferences and workshops on the subject, which are further highlighted in Chapters 6–8, suggest that the problems and issues that will be outlined are common across most practical projects, although obviously location and the players vary.

The Institute was set up in 1998–1999 with the aim of improving the quality of urban communities through the development and promotion of sustainable approaches to urban living. Thus, there was a focus on urban issues, and a deliberate aim to concentrate on practice and policy rather than theoretical debate. The Institute brought together expertise from environmental management, energy and building technologies, social science, design and public policy. This coverage, while extensive, still falls well short of the subject coverage required to do full justice to the subject of sustainable development. However, experience has shown that this is a common problem and most organisations studying the subject lack coverage in one or other of the main directions of social, economic or environmental. Thus, an instantly identifiable practical constraint was the difficulty of covering all aspects of the subject to the same degree within the bounds of expertise available and manageable numbers of staff.

An initial bias was identified due to the strength of the social strand of expertise. In practical terms, this has generally been an asset, since most other organisations interested in the research of sustainable development have an economic or environmental bias, allowing Institute staff to add value in partnership with others. Institute projects have had a range of initial objectives and desired end-goals. Many were not fully cross-disciplinary in nature, relying in part on a client's interpretation of sustainable development, itself a cause of irregularity.

Thus, the Institute's work in general allows identification of a number of important drivers and constraints common to many practical case studies throughout the sustainable development debate. A key driver was in choosing to concentrate on one specific area of the subject, the urban environment. The constraints included the practical constraints of subject coverage, the need to deal with the bias that comes from the balance induced by the initial strengths and weaknesses within the subject team, and the problem of dealing with the irregularities of client interpretation.

Projects have looked at subjects with a wide diversity of scales – from pan-European studies looking at common lessons across Europe through to local communities establishing their own approaches and goals for specific angles of sustainable development. This raises another important factor of the effect of scale and the appropriateness of the scale for each particular project.

Key to the Institute's success, however, is the need to find common lessons from the work if a common approach is to be developed. A number of examples of Institute work are therefore studied in this section to ascertain what are classified as sustainable development projects. This allows analysis of the trade-offs, constraints and compromises which need to be taken into account when addressing work in this area. Another important consideration is whether the main focus of much of the work at project level is biased in some manner specifically toward social, economic or environmental issues.

Many of the Institute's projects are easily identifiable as having a direct impact on the main theme for the Institute, sustainable development in the urban environment. The three examples below, however, were deliberately chosen from a wider list because they are diverse and have indirect rather than direct impact. It therefore takes more careful thought to analyse how they relate to the overall Institute objective and to the sustainable development debate in general.

The regional approach

The Regional Round Table on Sustainable Development is a regional partnership of policy-makers, business and various interest groups in the north-east of England (Sustainability North East 2001). It first developed as a discussion group to debate the merits of sustainable development in the region in 1998 at a time when there were few organisations operating at a regional level. It is now acknowledged that, while useful in initiating a debate, it became a talking shop in its early form.

In an effort to energise the Round Table to become more proactive than its previous largely reactive role, the Institute was asked to work with others to create a more clearly defined role for its future development. At the same time national government in the UK, which had been developing a regional agenda on sustainable development, asked existing and newly formed regional-level bodies to develop a sustainable development framework for each region. The Round Table became a natural vehicle for delivery of this work. Thus, the Round Table had both a past legacy and a desired role, albeit created by national government rather than by the region itself.

The domination of policy-makers and environmentalists on the original Round Table was a reflection of earlier views on sustainable development where the primary focus was on environmental issues and the public sector in the region was being driven by national government to take account of the subject in policy. The second attempt at the Round Table tried to redress this bias through a better balance of participants, although it continued to be a difficulty simply because the environmentalists provided much needed drive and energy while the policy-makers had the budgets to develop the work. Despite this, the main drive was to guide all strategies developed by regional bodies to become properly cross-disciplinary even where the main purpose of the individual strategies was to look at specific issues such as housing, economic policy or even culture.

It is debatable whether, in its original formation, the Round Table was at a sufficient scale to look properly at regional policy issues in sustainable development. Of the original main participant organisations, only two had a scope of responsibility that was truly regional. By the second phase, however, this had increased significantly as the region developed regional political structures, regional health frameworks and regional economic development bodies.

Analysis
The client's collective definition of sustainable development, the starting point for the project, had a strong cross-disciplinary theme, although the constituent organisations formed to look at the subject had clear initial strengths and weaknesses in each of the three key directions of economic, social and environmental coverage. In this respect the Institute had a role in adding to the balance of the coverage, although it was clear that further widening of participation was required.

At a regional level in a forum such as described above, political consideration is a key factor. Specific policy-makers, their budgets and key national agendas will continue to be key drivers in the debate, although the participants' perception of scale affects their ability to commit themselves to action and thus key players continue to remain uninvolved.

The regional level of governance is an important platform in the European debate on global v. local responsibilities (Denton 1981). It is seen as the ideal platform to feed a global and trans-national message down to more local levels, which includes local urban environments, from where action can then be taken locally. Thus, the scale was appropriate if measured against EU and UK guidance, a primary consideration for future policy in the region.

Two important concluding questions remain for the future development of the Round Table: can a truly balanced forum ever exist and, if it covers all topics fairly, can it reach beyond agreeing the lowest common denominator and develop new insights which lead to better and improved solutions? The work of the Round Table is still on-going and only time will tell if these questions can be successfully addressed.

Local approaches

Community Balance was a much smaller project where the objective was the support of a small team studying the use of environmental good practice as a catalyst for work on social inclusion in a deprived semi-rural area (Richardson 2001). A previous scheme at the same location had conducted similar work on a much grander scale using a 250-acre purpose-built facility, but had gone into receivership. This highlighted the difficulties of working in a relatively deprived location with a fairly innovative approach, a location unlikely to receive the attention and support needed to start such a venture.

A scaled-down version was developed by some of the original core staff and the Institute, and received funding as a transitional project, with the objective being to develop a stable succession project. Importantly, the core staff were very committed to the twin ideas of environmental good practice and social inclusion, and developed the basic idea into a series of participatory initiatives and events with various groups from the local community. On completion of the transition phase of the work it was then transferred to a charitable body with a local board of trustees.

Analysis
From a project point of view, the Institute would view the achievement of succession as a successful outcome in sustainable development. The Community Balance model is one of many

projects aimed at small-scale improvement and, although it was conducted in a semi-rural setting, the lessons are directly transferable to other areas needing regeneration, many of which are often urban.

Addressing the earlier common questions of the project, however, would reveal failings and gaps. The project had the luxury of ignoring the issue of economic sustainability, aided as it was by charity funding for a relatively secure 4 years. It was also probably less rounded than would suit a true definition of sustainable development.

The scale of the project remains, with hindsight, appropriate given the goals of local progress and participation. However, it was much smaller than the previous project and thus suffered little from the need to address the inevitable trade-offs and compromises which are a part of sustainable development. Like the Round Table project, there was the legacy of a past, failed project hanging over the new scheme, which inevitably skewed some decisions.

Overall, therefore, the lessons for global sustainable development in general were limited to showing the importance of local ownership of such projects and concentrating on what works on the ground.

Transnational projects

SUSPLAN was a large European-funded project involving three local government–university partnerships across Denmark, the Netherlands and the UK, looking at how attitudes to sustainable development impact upon urban and rural planning regimes (Porter *et al.* 2001). The team was looking for common lessons across northern European locations, an area of the globe that should produce many similarities which should lead to a common approach.

However, planning regimes are different in each country, and each country team had to specialise in issues of different scales and types. In the UK, planning at an urban level was studied since at the time of project inception there was no regional planning regime. In Holland the team looked at a new experimental level of planning where there was joint planning across two regions. In Denmark the setting was at county level, the apex of planning decision-making for many aspects of life in rural Denmark. The obvious conclusion drawn was that planning regimes are very different across the countries and therefore impose very different constraints on decisions and how decisions can be made. Importantly, experience (particularly bad experience) rather than theory dominates across the three countries, as the legacies of past failures loom large in planning and appear to have a disproportionate effect.

Trade-offs and compromises are a strong feature of the systems, and inevitably result in a complex web of influences, as Fig. 5.1 shows for the UK. The trade-off between complication and providing a platform for a range of views to be delivered leads to both gaps and overcomplication.

The project quickly revealed that no one country had a monopoly of best practice, and a healthy situation of trial and error occurs across the three. Again, the legacy of past failure rather than current practice was a prominent factor in the pre-conceived ideas of the teams. For example, the UK team were convinced that consultation was a weakness in the UK system and that Dutch and Danish teams could bring advice and guidance. In the event, the large amount of consultation that goes into UK planning was seldom matched by similar levels of participation elsewhere.

Analysis

The message of the project for global sustainable development was that, while common tools for the various stages of planning and a common overall framework could be devised for use transnationally across the very different situations in the three countries, the outcomes, goals or objectives of each system required local decisions on local priorities.

The problems that arose from differing initial priorities, differences of scale, differing interpretations of the meaning of sustainable development and a desire to look at different sections of the planning regime were, in one sense, put to one side initially since partner countries concentrated on their own systems and strengths. However, the outcome, given this starting position, was a remarkably strong model for future study (see Chapter 6).

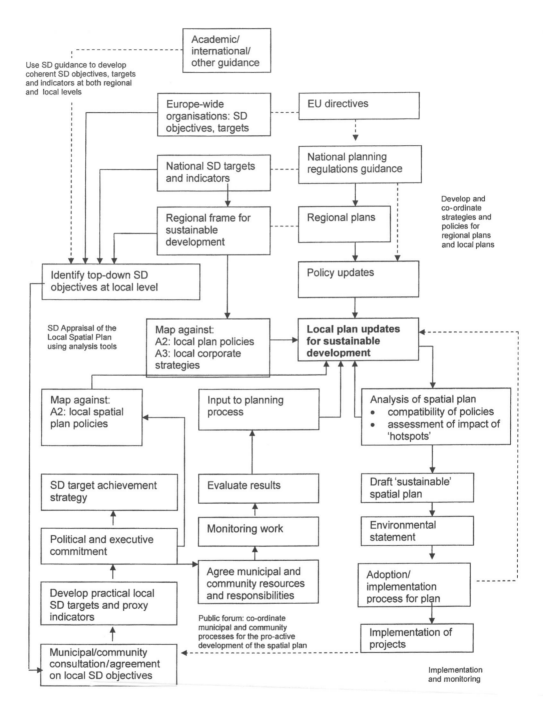

Fig. 5.1   The integration of sustainable development in planning (Porter 2000).

5.2   Keys to success

All three case studies approached global v. local themes in different ways. Each experienced different coverage constraints, had differing initial problems with strengths and weaknesses, coped with differing client perceptions of sustainable development and addressed differences arising from scale. Despite the diversity there are common lessons to draw, with the most critical of these being the importance of appropriate scale for each project and its participants, and the need for appropriate goals which take account of local circumstance and ownership of the issue, problem or task. Avoiding unnecessary complexity is another factor that appears to influence success.

However, none of these factors is easy to assess because they appear to involve a mix of fact and perception; direct, easily identifiable influences together with local perception of what is wrong and what is achievable, which may not match. This brings us back to the lessons highlighted in Fig. 1.8 and suggests the need for a model that takes account of part fact, part politic, part unknown.

Scale, which is further analysed in this and the following chapter, is a critical issue simply because there is no right and wrong scale for dealing with sustainable development. There are, however, clearly scales of appropriateness which are influenced by a number of variables. To show this further, Table 5.1 highlights the problems that arise with Institute projects from use of measures of success which might be considered at regional or project level. A key question concentrates on the measurement of success and trying to define a suitable currency of measurement (this is further discussed later in the chapter). The units of measurement considered are those that often find their way into debates on sustainable development, but they show themselves to be largely unsuited to smaller project level. The table raises issues of information availability, complexity and levels of acceptability. Other difficulties surround the balances that are required but which vary with each situation, making choice difficult.

The clearest lesson from the case studies is the explanation of the many barriers to the one-size-fits-all approach to defining sustainable development. This explains why so many practitioners see the need to redefine sustainable development to suit the circumstances of each new project, which, although justifiable in the

Table 5.1 Measuring success at Institute project level.

Dimension	Possible measures of success	Information availability	Acceptability
Overall	Client satisfaction?	Available	Qualitative rather than quantitative?
	Menu of indicators	Not always available	Examined in Table 6.2
Environmental	Index of Sustainable Economic Welfare (ISEW) (Section 2.4)	Too crude at project level	Not applicable (but fashionable in EU circles) Still viewed as academic
	Ecofootprints (Table 3.3)	Possibly available	A very narrow measure of success by itself
	Material waste (Table 3.2)	Possibly available	A very narrow measure of success by itself
Social	GDP (Chapter 2)	Too crude at this level	Not applicable (but accepted at national levels)
	Job gain	Possibly available	A very narrow measure of success by itself
	Deprivation index	Possibly available	A very narrow measure of success by itself
	'Feel-good' factors	Available	Largely untried but signs of favour within EU
Economic	Cost-benefit (Chapter 2)	Available	Assessment depends on choice of size of area affected
	GDP improvement	Too crude at this level	Not applicable
	Whole life cost (Chapter 4)	Available	Largely untried but signs of favour within UK
	Cash-flow generation	Available	Does it fully cover the benefits?

current debate, is wasteful and counter-productive long-term if it can be avoided.

The define–redefine dilemma forced more debate within the Institute and thinking on what standardisation was possible. Some of this work has already been discussed in the first chapter and this suggested that it is doubtful whether 'balanced' outcomes are always desirable. However, further work was required to re-examine possible simplification and whether this can be mapped mathematically. This is explored in the following sections.

A second area of investigation concentrated on the study of inter-disciplinary work and the numerous barriers that become obvious. This included addressing the inevitable value judgements required in a sustainable development evaluation process and the inclusion of democratic ownership which is critical. This work is further considered in Chapters 6 and 7.

5.3 Simplification of the way forward: defining the process

The standardisation of sustainable development takes up the arguments of the first chapter and the question of whether sustainable development is a concept that represents starting-point, process or end-goal. This and the next section return to the study of process and end-goal, both of which generally interest practitioners more than theoreticians. The starting-point arguments will be revisited in the next chapter.

Standardisation can imply simplification of the issue, and this itself can cause problems. As noted before, the current media debate often centres on an oversimplified debate between two extremes, the 'weak' versus 'strong' green argument (Department of Environment, Transport and Regions 1999, European Commission 1996). Priorities that are set based on a political choice between two extremes are unlikely to be subjective or optimal, particularly in a subject where 'evidence' will take some time to catch up with 'facts'. As suggested in Chapter 1, it is possible that there is an envelope of solutions for most development opportunities. This makes it harder to define a universal answer and leads to the simplified choice often put forward. Some of the various simplifications are, however, useful short-term and aid clarification of starting-point, process or end-goals although they create problems long-term in defining a consistent theory and principles. At a two-dimensional level these simplifications have included:

- *Socio-economic (human-centred) v. environmental (nature-centred)* – as illustrated in Chapters 2–4
- *Optimist v. pessimist* – raised by Pearce *et al.* (1990); many of those in favour of the status quo are optimists who believe that humanity will deal with the problems in the fullness of time. The pessimists seek to prevent damage before it occurs.
- *Right wing v. left wing* – a favoured explanation of the media.

However, while these allow clarification of some of the issues, they provide little help in achieving a consensus on the process of moving forward. At the next level, there are three-dimensional representations of the debate. The combination of social, economic and environmental forces has produced a variety of simplified illustrations to explain the concept (Fig. 5.2). These suggest three main variables, with the nature of the interaction between the three dimensions represented in a number of manners; as the three legs of a stool, as three overlapping circles, as three interlocked circles or as a triangle.

The geometrical illustration can often be described mathematically. Does mathematical modelling assist development of a process map? There are many who see this type of approach as a possible way forward and this has already been noted in Chapter 1 where the case study on the Department of Environment, Transport and Regions (1999) checklist revealed that a variety of interactive models have been considered, the most famous being a Pressure–State–Response model. It is accepted, however, that many of these are complex and require very specialist input and expertise, an issue discussed further in Chapter 7.

In Chapter 1 the use of the continuum of a triangle as a method of tracking the process of sustainable development was discussed (Fig. 1.5). It was suggested that, as with all mathematical simplifications, the results developed are less than ideal, but it does allow examination of the problems and advantages of mathematical modelling. This is now examined further.

The two-dimensional representation shows the movement from a pre-project set of social, environmental and economic circumstances to a post-project one. It is useful since it raises the notion of loss and gain in the three directions, but it runs the risk that the triangle's centre of gravity will be seen as the ideal end-point, the perfect balance between economic, social and environmental. More importantly, scale and displacement become more obvious problems since benefit and loss can also occur beyond the boundary set for the triangle.

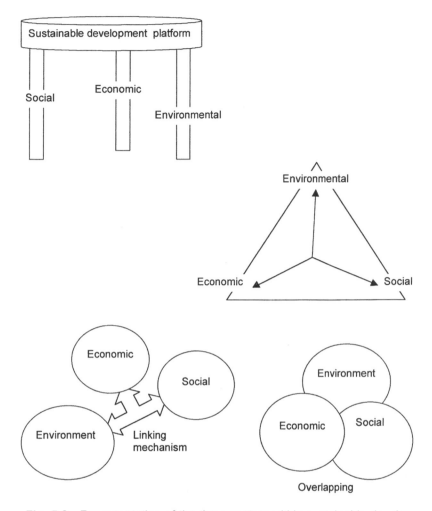

Fig. 5.2 Representation of the three vectors within sustainable development.

A simple method of expanding the two-dimensional mapping is the idea of two superimposed triangular planes to deal with the local effects and the global effects (Fig. 5.3). This allows us to deal specifically with scale and displacement effects by mapping out-of-plane effects onto another plane. This, again, is a gross simplification for analytical purposes, but it is a useful representation, allowing examination of scale and displacement.

A good test case for this is Girardet's 'mango' example (Girardet 2000). Girardet quotes the example of mango produced in the Philippines and flown to Europe for consumption as a classic example of unsustainable development, because the energy

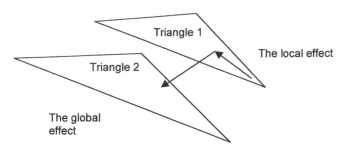

Fig. 5.3 The two triangle theory.

contained within the mango is vastly outweighed by the energy consumed in the journey to market. To better understand why the situation occurs, the top triangle explains how at the local level it is a win–win situation, economic and social benefit arising in the Philippines from selling the fruit with little or no environmental degradation locally. In parallel, there is economic benefit in Europe from increased choice and an affordable food.

Triangle 1, based on the situation in the Philippines, shows the use of a natural resource in a renewable sustainable manner with little or no movement across the triangle. A similar triangle could be drawn for the European end of the trade. However, the travel involved and the energy consumed in transporting the mango by aeroplane causes more significant damage at the global scale. Since the effect at both ends is an apparent win–win it is difficult to take action to stop it. Stopping the mango trade would cause negative impact in the Philippines with job loss and deprive European consumers of choice. The problem of energy use in-between is invisible to both sides. Dropping down to the global triangle (triangle 2) would allow us to show the environmental degradation occurring elsewhere, a classic displacement problem.

A conclusion from this type of representation is that it is still too problematic to be robust. The effort involved in developing a complete picture is significant, making it beneficial only for large projects. This may rule out individual lifestyle decisions, even although collectively they will cause damage. The method is still mapping rather than optimising decisions. It does, however, further highlight the gap between environmental effects and socio-economic effects, possibly promoting the case for a two-variable model rather than three variables. This, however, would need to avoid the current tendency to split towards two extremes.

5.4   The constraints and barriers to standardising the process

The constraints and barriers to establishing a standardised process based on analysis derived using a three-variable model are therefore considerable. However, there are a number of simple steps needed for good practice in evaluating development. It is now clearer that the simple steps developed in Fig. 1.3 can now be expanded into a more detailed eight straightforward steps (Fig. 5.4):

Defined clarity and balance

      Initial assumptions?

            Appropriate scale?

          Methodology

              Optimise solutions

                  Normalise language

                      Decision sensitivity analysis

                          Scrutinise measures of success

Fig. 5.4   The DIAMONDS steps to process development.

(1) *Define aims* – clarity is very important and must be a top priority. The supplementary question to add to this derived from this chapter is what sort of balance is needed in each individual project in terms of economic, social and environmental cost-benefit? A balance of least damage and most added benefit?

(2) *Scrutinise initial assumptions* – again, the emphasis is on widening the search for options and analysis to look for bias and gaps in the initial assumptions so that they can be eliminated where possible.

(3) *Establish constraints* – the key requirement to add to normal best practice would be the need for appropriate scale. This is often considered in large-scale projects but can be problematical in smaller-scale projects. It is, however, critical if a full solution is to be investigated, and is part of the argument behind ecological footprints.

(4) *Common methodology* – this is still the big unknown; such a methodology needs to make reference to the principles of sustainable development, to the split between fact, political choice and unknown, and to accept the fact that all players do not see immediate benefit.

(5) *Minimise/maximise key parameters* – the driver behind the use of indicators in evaluation of developments since indicators are, of course, a proxy for the key parameters. However, it is often difficult to see how they relate to the fundamental principles, and there is seldom any attempt to minimise or maximise a solution (on the grounds that it is too difficult to measure).

(6) *Common language* – this is a problem that should have been eliminated but still continues to cause problems. It clearly requires special measures and perhaps new processes to eliminate the problems of cross-disciplinary discussion.

(7) *Sensitivity analysis* – such analysis receives little or no attention, but, given the political nature of many of the decisions involved in sustainability issues, it would clearly be a useful step if we knew how much leeway was available for any given solution.

(8) *Results* – what are the benefits and how are they measured? Currently the answer is often either financially or with indicators that are often not numerically summated. There are good reasons for this given the difficulty in measuring the indicators, the lack of a common language and the abstract nature of some indicators. However, the avoidance of any numerical approach leaves a wholly subjective approach, which can be suspect.

The response to the above of many good project managers would be to argue that best practice dictates are the same (HM Treasury 1995), so what is different? Certainly implementation of the above steps is likely to raise the cost of developing and evaluating proposals and projects. However, best practice continues to be an elusive goal. Few proposals can claim to fully follow best practice. The above list adds small but important pieces to the jigsaw and the slight changes in emphasis from normal best practice can make a significant difference.

The key gaps that prevent the above from being a workable check-list are three-fold: the common methodology (which will be studied further in Chapter 6), means to normalise language (which will be studied in Chapter 7) and developing widely accepted measures of success. The development of widely accepted measures of success is a critical problem for all stages of the process since the availability of acceptable, measurable targets, indicators or proxies would allow clear definition of aims, assumptions, methodology and measures of progress. Thus, much depends on finding an acceptable, all-embracing currency for the end-goal, whether it is through indicators, new composite measures or some other form of proxy.

The arguments of Chapters 2–4 indicate that any currency that is too deeply associated with one of the three elements of sustainable development (social, economic or environmental) may have difficulty in seeking universal support. Table 5.2 returns to the currencies shown earlier in Table 5.1 and looks at the issues that arise at a more general level. In addition to the problems of association it indicates the common issue that short-term measures are often different from the best long-term measures (Hawken *et al.* 1999). Breaking these into the three elements:

- *Economy* – at national or even regional level the measure would be GDP measured in monetary terms, at project level it would be cost–benefit measured in monetary terms and for most other applications it would be monetary gain. The advantages and disadvantages of this have already been discussed in Chapter 2, but clearly these measures have some difficulties in crossing over into environmental and social aims and, more importantly, long-term planning.
- *Environmental* – attempts have been made (Jackson *et al.* 1997) to produce a monetary measure of environmental damage. Although useful, it is still not seen as a reliable alternative to GDP. More importantly, factors such as space, energy and diversity are important components in environmental progress, but are not measurable in monetary terms. The eco-footprint for cities or physical developments, for example, is useful but limited to space.
- *Social* – again, as illustrated in Chapter 2, social issues have been measured through, for example, GDP, although in general it is recognised that more abstract factors are more important if difficult to measure, the 'feel-good' factor being a good example.

There are many accepted proxies, such as confidence surveys, hybrid deprivation indices or the human development indices, which can quantify qualitative measures in broadly acceptable terms, although they are seldom seen as equivalent or equal to economic or environmental 'facts'. The important lesson is there is no clear-cut proxy for each of the three dimensions and therefore not three common starting points or currencies. The best choice depends on each development, its scale and diversity. This leaves

Table 5.2 The possible currencies and their shortfalls.

Dimension	Measure	Short- or long-term bias to analysis	Issues for universal acceptance
Environmental	Monetary	Similar to GDP	No universally accepted method as yet
	Ecofootprints	An immediate situational measure	Factual but narrow?
	Raw material wasted	An immediate situational measure	Factual but narrow?
	Energy	An immediate situational measure	Factual but narrow?
Social	Monetary	See note on GDP below	More directly linked to economic change
	Job gain	Is open to short-term influence	Can show rapid but unsustainable variation
	Deprivation index type	Can map long-term change	Hybrid indicator type approach
	'Feel-good' type	Very often open to short-term influence	Can show rapid but unsustainable variation
Economic	Monetary (cost–benefit)	Three- to five-year business plan orientated	Complications of analysis
	GDP improvement	Can map long-term change	Criticisms from social and environmental
	Cash-flow generated	Critical but short-term	Similar problems to that with GDP
	Life cycle cost	Long-term?	Still lacks data and methodologies
Human	Time	Short?	Driven purely by busy businessmen?

the way open for the growing indicator industry to provide a tailored, short-term solution.

Table 5.2 reflects discussion at a recent conference on the Sustainable Information Society. Rifkind (2001) suggests that much of business and the economy is moving from a system that promotes profit and ownership to one where access and time are the critical factors. This then implies that the main currency is switching from a monetary measure to one of time, an interesting concept. However, while this belief may have a certain logic for top business people who are seldom short of money but starved of time, it is not a universal driver. Hawken *et al.* (1999), for example, point out that this is the first era with unlimited labour supply but limits to natural resources, whereas in the past limited labour supply and unlimited natural resources have been the driver for progress through agrarian and industrial periods.

A goal of the sustainable development debate must therefore be to seek three common starting points, which are preferably long-term in analysis, and an agreement on the process to quantify the movement in the three directions. This, however, will not be easy, as Table 5.2 shows, although it would represent a step forward from the current system since questionable value judgements occur at all stages throughout the current non-standard processes. Such a system with such a currency could then leave the 'political' value judgement to the final stage, seeking agreement on (1) fact, (2) specific weighting on which factor is most important for a particular scheme – environmental, economic or social gain or the balance between the three, and then (3) specific political choices.

Table 5.3 highlights some work by an expert team for the EU (DG-Environment 2000) which developed some sustainable development indicators for local use within European municipalities. Unlike many other efforts in this field this project work has considered the principles (see the discussion at the start of Chapter 6). However, it is clear that even in this simple system political choice plays its part, while the global link and measurability of the indicators cause many difficulties. Only those factors that are truly within the reach of citizen and municipality can be influenced.

5.5 Summary

The study of practice and the experimental work at the Institute point to many issues that are important to the sustainable development debate but are often underlying rather than immediately

Table 5.3 European common indicators. Crosses indicate coverage of the subject.

	Local effect	Global effect	Influencing factors (key players)	Measurability
Citizen satisfaction of local services	x	?	Citizen and municipality	Consistency of measurement?
Local contribution to global climate	x	x	Cars and industry?	Availability of data?
Local mobility and transportation	x	x	Citizen and municipality?	Complexity of data?
Availability of green space	x	x	Citizen and municipality?	OK
Quality of outdoor air	x	?	Climate, cars and industry?	Location specific?
Children's journey to school	x	?	Citizen and municipality?	OK
Sustainable management of local services	x	x	National priorities?	Consistency of definition?
Noise pollution	x	?	Citizen and municipality?	Location specific?
Sustainable land use	x	?	Municipality?	Complexity of data?
Eco-product promotion	x	x	Fashion and industry?	Availability of data?

recognised. Issues that have been identified as 'factual' are further discussed in Chapter 6, while those identified as linked to individual perception are further examined in Chapter 7. These include effects of attitudes and approaches to learning and the legacy of preconceptions and perception. Some interesting conclusions can be drawn. Some of these factors can, at first sight, look irrelevant to the end solution of unravelling clear definitions for sustainable development. However, the whole is the sum of many smaller factors and these factors can have critical effects on the whole:

- Democracy – ownership is critical in an issue with no clear rights and wrongs. Consensus and the capture of all good ideas relies on participation which in turn needs a desire to run by democratic means (a subject of interest in Chapter 6).
- Sorting out the language – the current define–re-define has become an easily exploited area for new language to thrive, which in turn makes understanding more difficult. In presentations to lay people it becomes very obvious that the experts have lost the audience, an issue addressed again in Chapter 7.
- Clear aims for joined-up thinking – much of the early work at the Institute concentrated on achieving a balanced team, assuming that the best end-goal was a balanced outcome from this balanced team. Hindsight would suggest that a balanced outcome is not necessarily the most desirable outcome. Nevertheless, natural bias must be acknowledged and some effort made to address it.
- Optimised solutions – efficiency takes many forms, but it lies at the heart of many sustainability problems and must be acknowledged, as will be noted in Chapter 6.
- Displacement – the displacement of problems to another dimension, jurisdiction, etc. is not a sustainable solution, and this is addressed again in Chapter 6.

Practice provides the evidence from which the process of sustainable development evaluation can be tested and mapped. Some brief experimental work in mapping reveals some further interesting insights: it reinforces the importance of scale and displacement, and it highlights the dangers of oversimplification and suggests that there may be a range of solutions, though not necessarily the extremes often advocated by 'experts'.

It is important to remember throughout that a decision on sustainable development is based on part hard fact, part unknown or unmeasured fact and an element of political priority, as was first noted in Chapter 1. Thus a balance is needed between participation and best choice, again a lesson that can be drawn from current best practice guidance in the public sector.

Sustainable development should involve improvement and it needs a measure for this that embraces more than economic consideration. The current hybrid measures of improvement are not sufficiently developed. Alternatives, which involve the separation of the social, environmental and economic into three (or possibly two) factors and their separate measurement, are equally imprecise at present.

The work in this chapter has again highlighted the need for further development of a process model for sustainable development evaluation, which must be robust enough to be employed either post-event or in the pre-planning stages. An acceptable, all-embracing currency to measure success remains elusive, and it is likely that there will continue to be a reliance on the use of indicators as a proxy.

Sustainable development as a subject should provide the medium of understandable universal principles. Practice-based work has progressed quickly, and highlights the complexity of the current definition of sustainable development and its obvious defects to practitioners. The complexity arises from the interdisciplinary nature and the differing approaches of experts in each field. Progress on clearer understanding to underpin a coherent approach remains slow, and each new piece of practical work brings a redefinition of the end-goal, an unsatisfactory but understandable situation given the current theory vacuum.

6 Missing Elements in the Debate

6.1 Missing steps in current practice

The previous chapters have studied how the subject of sustainable development incorporates a wide combination of disciplines related to a loosely defined set of principles. This mix can be interpreted in a variety of ways. Some experts suggest a need to maintain the status quo with some minor adjustments. Others see the need for radical transformation to humanity and its interaction with planet earth. A few still have more mixed views or advocate some particular change in specific areas of society, the economy or the approach to the environment.

In the second part of the last chapter there was a focus on the process of evaluating sustainable development. It was noted that this still presents difficulties. The first part of this chapter will therefore concentrate on the principles, highlighted in Chapter 1, and how they feed directly into sets of indicators without an apparent consistent process. In the second part of the chapter the emphasis will shift to well-hidden elements of the debate which may provide clues on why there is so diverse a set of views associated with the subject.

There is little or no standardisation of the process of evaluation for sustainable development, beyond a general acceptance of indicators as a suitable proxy. The indicators employed can vary enormously, as Fig. 1.3 showed. Thus, a standard model of the assessment process, without a prescriptive end product, is an essential in the current sustainable development debate.

The only prerequisite for any end-goal which results is that it must clearly show one form of payback as preferable to others, so that any initial bias can be acknowledged. The lesson from the SUSPLAN project highlighted in Chapter 5 is that all players need to see a tangible, comparable payback if they are to become willing participants. Tangible proxies, such as indicators, need to be employed in a systematic manner both as part of the evaluation and in describing the end-goal.

It was noted in Chapter 1 that the principles and language vary, ensuring difficulty in comparing and contrasting concepts and issues. Table 6.1 summarises a sample of the principles behind the definitions studied in Chapter 1. At the level of 'principle', the constant factors which appear across all the schools are the inclusion of reference to a social improvement requirement, a durability (long-lasting) element and an environmental requirement, although each emphasised to different degrees. These three basic elements are needed in seeking any solution to a sustainable development problem. It needs to be noted, however, that much of the detail that follows in these references also includes an economic element, a possible acknowledgement of the current working system and its three principal defects.

Table 6.1 Key principles drawn from a sample of definitions.

Source	Principles
UK government (DETR 1999)	A better quality of life
European Union (CEC 2001)	User pays and polluter pays
World Bank (2001a)	Balance across three basic elements
Pearce *et al.* (1990)	Futurity, equity and environment
Hopwood *et al.* (2001)	Futurity, equity and participation

It is useful to move down the hierarchy from principle to element and below in search of a common currency (Fig. 6.1). Many studies of the subject have attempted to map the complex variety of building blocks within sustainable development in terms of indicators, as was noted in Chapter 5. This is viewed as a suitable simplification for practical purposes, although it often produces flawed indicators. It is difficult to see any consistency in the move through the hierarchy from principle through to establishing what elements are needed and a suitable set of indicators. To make matters worse it is also hard to see how indicator theory, which requires information availability, acceptability and neutrality for best practice, has been applied to the final choice.

Table 6.2 shows a sample of indicators in current use and compares them with the requirements of best-practice indicator theory. Not surprisingly, this brief survey shows that there are many failings in terms of neutrality, acceptability and information availability. As with Fig. 1.3 earlier, it is interesting to note the range of numbers of indicators used. Consistency and complexity are clearly primary problems. Other factors such as the relative weightings attached to each indicator and methods of combining

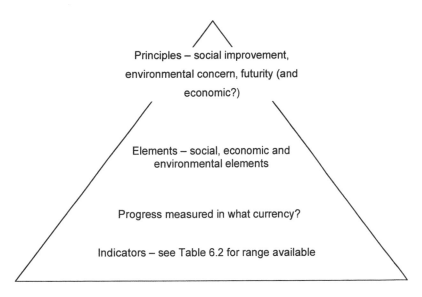

Fig. 6.1 The hierarchy of sustainable development factors.

them when there are overlapping effects all leave many lists open to bias, although it is often inadvertent.

Work at the Institute suggests that there may be three options for evaluating sustainable development which may form a basis for linking indicators to the principles (Table 6.3). The methods, briefly outlined in the second column, need to pay special attention in their formulation to indicator best practice.

Is this over-analysis of an already complex subject? Many practitioners hang on to indicator theory as the only way forward for evaluation of sustainable development, but if the indicators are missing vital areas of subject coverage then their value must be questioned.

With all this effort it could be argued that there is little new to discover, or that trying to add further factors to the equation will only further complicate the subject. Surely the work on indicators has been broken down sufficiently to identify all the building blocks and all the important elements necessary for a full debate? However, the previous chapters have revealed important factors that are often just footnotes in many mainstream documents on the subject.

Thus it is important that these factors are highlighted, reviewed and prioritised if the debate on sustainable development is to move forward without significant gaps. The next section therefore looks at seven factors that increasingly appear to be significant

Table 6.2 A sample of sustainable development indicator lists and models in existence.

Source	No. of indicators	Neutrality?	Acceptability	Information availability
UK government (DETR 1999)	160	Wide coverage, but is it neutral?	Origin of indicators is unknown	Acknowledged problem of no information with some of the areas of coverage
Seattle urban area (Best *et al.* 1998)	40	Wide coverage, but is it neutral?	Civic panel choice	Ten indicators have insufficient data
World Bank (World Bank 2001a)	15	3 × 5 principles, 'balanced'?	Conceptual – origin unknown	Guiding framework rather than hard indicators
EU – sustainable cities project (DG – Environment 2000)	10	Coverage? New? Neutral?	For local use but expert team derived	New indicators – little information available
EU – European Foundation (Mega and Pedersen 1998)	16	Mixed? Coverage? Neutral?	For local use but expert team derived	Composite indicators – complex assumptions need challenging?
Arup – SpeAR (Arup 2001)	20	4 × 5 principles, 'balanced'?	Expert derived – is this OK?	Yes – probably geared to information rather than need
ESI (Esty *et al.* 2001)	60	Environmental bias	Expert derived – is this OK?	Yes – but complex assumptions need challenging?
OECD (OECD 2001)	50	Environmental bias	Expert derived – is this OK?	Yes?

Table 6.3 Three 'principled' options for evaluating sustainable development.

Principles	Methods	Outcomes
Equity, participation and futurity	A series of interviews with stakeholders focusing on present, desired and potential contribution (de Boer and de Roo 2001)	Often presents a range of views and some good ideas
Social, environmental and economic balance	Uses known checklists (see Table 6.2 examples) as base cross-reference	Output is commentary but relies on impartiality of checklist and assessor
Inclusive, joined-up approach	Cross refers to other 'local' relevant documents top-down and laterally (see Fig. 5.1 example)	Often desktop study with strong local content

oversights in current studies. Not all of them are 'lost' as the chapter heading suggests, but they are certainly well hidden, even where they may be implied through inclusion of other elements in the equation.

6.2 The 'lost' factors

The identification of gaps within the current debate and the analysis of their implications are crucial to progress. Some of the gaps help to explain some of the tangential positions of the players – 'one man's meat is another man's poison' in both language and substance.

The seven factors are listed in Table 6.4. It is interesting to note that some of the seven factors are actually well discussed in certain corners of the debate but have failed in the past to make it onto the main agenda of the majority of organisations working on sustainable development. The following section provides a more detailed explanation of each, a hypothesis on why the factor is important, its effect and some observations on why the factor has been ignored, missed or hidden.

The first three factors in the table are flagged as priorities which have been missed by much of the key literature in the sustainable development discourse to date. The second four are important but tend to have been captured by one school or another and ignored

Table 6.4   The seven 'lost' factors.

Factor	Why is it important?	Hidden or missing?
(1) Scale	Primary importance to economic (economies of scale), social and environmental (natural levels of population retention) capital – underpins all other factors	Missing
(2) Displacement	Primary factor but often beyond the bounds of theory (the result of the action typically occurs beyond the boundary of the area of study) – underpins importance of 1, 5, 6 and 7	Missing
(3) Value judgement	Primary factor since it fills the gaps in evidence but seldom acknowledged as a subjective judgement – with strong link to 4	Missing
(4) Democracy	Secondary – appears in some literature but not others; often tainted by association	Hidden
(5) Efficiency	Secondary – appears in some literature but not others; often tainted by association	Hidden
(6) Space	Secondary – appears difficult to quantify	Hidden
(7) Population	Secondary – appears in some literature but not others; often tainted by association	Hidden

by others. However, the issues need to be considered by all schools if arguments are to be credible. A further complication is that these factors seldom work in isolation and, as a consequence, can be hidden within other effects or lost in different language. This makes them difficult to analyse and helps to explain why they do not feature heavily in current debate. This, however, is no excuse for ignoring them as critical issues. The factors will therefore be individually addressed in this section.

6.2.1   Scale

Scale has always been an important aspect of development. Its effects have been noted throughout the earlier chapters. Finding the most appropriate scale for a development is critical, whether it be to maximise economic profit by maximising returns and minimising cost, or maximising social benefit and minimising environmental intrusion.

At the Institute, on-going project work at both community and pan-European level quickly exposed the difficulties of translating lessons from a very local level to an extra-national level. The conditions and priorities in an isolated rural market town in, for example, northern England seldom match the priorities of the nation as a whole. Berwick-upon-Tweed, a small, pleasant town of 27,000 people on the English border with Scotland, has suffered in the past few years from a series of decisions designed to promote sustainable development. It is also, in a sense, typical of many Europeans towns since half the urban population of the EU live in towns of less than 50,000 population (Ahti 2001).

Globalisation, seen as good sustainable development by the economists, has left it isolated geographically and losing jobs. Regionalisation, seen as good by the EU since it brings decision-making down to a more local level, has left it on the fringes between the regional centre at Newcastle and the Scottish capital in Edinburgh. Banks and schools have been removed because of the economies of scale that suggest insufficient customer or pupil bases. Moves to change agriculture to a more green, less intensive form have reduced income in the area.

As a rural town of 27,000 people, Berwick-upon-Tweed appears to have a scale that seldom matches the requirements of the policies being produced in central government or elsewhere, and the basic choices which are crucial to its future development are largely made elsewhere. However, it is a pleasant place which has many traits promoted as good for sustainable development. This would include a sense of community, a good environment and a stable but not very exciting economy.

The unfortunate conclusion is that towns like Berwick are unsustainable from a business and government decision-making point of view. They lack the economy of scale which has become a diktat in many circles. This, in turn, leads to the demise of social capital which adds to the difficulty of servicing them. This is a mismatch with human choice in the developed world which, given a choice, tends towards exactly this type of town as the ideal. The economic logic behind such decisions would lead to the removal or run-down of all towns that were not the 'right size' which effectively means removal of half the urban population of the EU.

In the developing world, by contrast, the drift of people from such towns to mega-cities with their economy of scale is viewed as a global problem that needs to be addressed since it leads in turn to the difficulties of rapidly growing cities which cause immense social and environmental suffering. The basis of the analysis that accompanies the emerging isolation and breakdown of such towns therefore needs to be challenged because it diverges from natural human opinion and it creates social and environmental problems if left unchecked.

In the UK, treasury guidance on project evaluation on scale effects is extensive (HM Treasury 1995) and, if followed to the letter of the guidance, provides a useful starting point for standardisation. Thus traditional economists are well aware of scale as an issue, although an issue is that their methods throw up solutions that tend towards only big (in economic terms) is beautiful.

There are many situations where, for example, the level of resources, the number of people involved and the environmental constraints are so obviously different that it makes it difficult to translate best practice. An obvious example was shown in Table 5.2 where currencies to measure success were highlighted. There is further examination of this subject in Chapter 7 since it causes fundamental difficulties for cross-disciplinary study and modelling of complex situations.

Scale is a major cause of the current view within sustainable development circles that practice will lead theory. The problems of scale makes standardisation of guidance difficult, and modelling (an important aspect of theory) thus becomes very problematic. Scale is a fundamental issue which has been missed rather than hidden in most sustainable development literature. The omission is difficult to explain, except in that scale may be viewed as an inherent factor within many of the other arguments. Within traditional economic circles there has been a strong belief that large is best (the economy of scale model). Within greener and more social circles there has been a counter-belief that small and local is best (Girardet 1999). Both claim that their view implies appropriateness when in fact they might be working at the extremes of an envelope.

Part of the problem is the difficulty in judging what is the right scale. One example of the problems encountered is that both economics and environmental science would each point to thresholds of acceptability which create discontinuity, i.e., that there are levels of population, pollution or behaviour (or whatever the factor may be) below which the environment or health or social structure or the economy can tolerate, but at a certain point it becomes unacceptable, too disruptive or harmful. This switching-point is difficult to map in theory and practice.

In other cases, there are issues where the most appropriate situation for input purposes (e.g., the best configuration for delivery of supplies and the most efficient working practice to a shopping mall) is not the most appropriate for output purposes (e.g., customer access, comfort and purpose). This is not a new problem arising solely from the need to consider sustainable development. Large organisations have been grappling with this type of complexity for many years trying to optimise the design of services, environments or projects to best serve a number of factors. It does, however, become more important if the principles of sustainable development are to be developed.

6.2.2 Displacement

Displacement is the phenomena where the solution of a problem at a particular time or location causes a problem in another location, dimension, jurisdiction or even time, i.e., the problem is moved on rather than solved. An oft-quoted example of this in the developed world is out-of-town developments such as shopping centres where it is often suggested that they do greater economic damage to established shopping areas than is offset by any new capacity generated. Socially, they involve people adjusting to the location rather than the location adjusting to the people. Environmentally, much has been written on the issue of increased traffic as a result of their existence. Greenhalgh *et al.* (2001) have suggested that one-third of regeneration successes in north-east England left voids elsewhere, and that generally out-of-town locations replaced services in city centre fringes or suburban locations, leading to more traffic and less access by anything other than the car.

Part of the problem is the difficulty in measuring the problem and providing a universal solution to the basic issue. In social science and regeneration circles, the chaining of events or actions has shown how a problem of, for example, empty shop-fronts has been displaced from one municipality with economic problems to a neighbouring one through inappropriate but well-meaning state intervention (Robson *et al.* 1998).

Unfortunately, many of these effects may be in different local jurisdictions so that the negative economic effect is felt in a different location to the positive. Thus, a development or solution involving displacement is clearly not a sustainable solution, since it involves moving the problem on rather than eliminating it. Once again, this is not a new problem and features in guidance from many government treasury departments or development banks on evaluation of public sector projects. Arguably, social scientists and regeneration specialists have tended to see this as a bigger priority than others since their desire is to see a specific problem such as homelessness solved rather than move it on. Pure economists often argue that moving the problem can still involve some benefit, if the new location can better cope with the problem.

Environmentalists appear to use different language for the same issue. In fact, the natural environment is the major recipient of displaced problems from social and economic issues. The problem, for example, of waste generated in urban areas is displaced to a land-fill site or to the sea, where it causes a different set of problems dealt with by a different set of experts using a different budget.

Displacement is key to sustainable development and will be addressed further in Chapter 8. As a concept it does not feature in most sustainable development literature, although it is inherent in many of the arguments, case studies and debate, principally under the disguise of the 'think global, act local'. While this is useful, it does not allow full recognition of the issues. It was noted, for example, in the last chapter that Girardet (1999) plots the path of a mango grown in the Philippines, transported by air and consumed in the UK as a displaced problem. However, the fact that it has beneficiaries at either end of the chain with the damage hidden in the transportation in the middle causes problems in recognition and raises difficulties in finding acceptable solutions.

It may be that *scale and displacement* – finding the appropriate scale and the need to avoid pushing the problem elsewhere – are two aspects of the same problem and can therefore be brought together. However, the mango problem shows that solutions are not straightforward. Eliminating any food produce that travels by air would have the net effect of reduced socio-economic capital at both ends of the chain, which may be more than the environmental capital saved.

6.2.3   Value judgements

The effect of value judgements was first highlighted in Chapter 1 (see Fig. 1.8) and is inherent in choosing the best combination of scientific or factual versus the political element of decision-making. It is clear from the previous chapters and from work on the ground that the easy decisions based on scientific fact are outnumbered by those where the facts are not quite so clear-cut.

Dealing with problems by breaking them down is a well-accepted manner of scientific approach and leads to an acceptance of single-discipline expertise, and its particular choice of methodology for collecting and analysing evidence together with its choice of currency. However, there are many subjects such as urban or spatial planning where the approach to the subject comes from a holistic approach with non-quantifiable solutions forming the basis of some models. In these areas, value judgements are accepted as part of the system or method of collecting and analysing evidence. Planning experts, for example, are among the few who acknowledge the importance of value judgements in sustainable development, recognising that few judgements can be based on fact alone and even technical subjects have an element of value judgement, even if it is only based on prioritising the time devoted to it (Healey 2001, Hansen 2001).

The value judgement factor seldom features in mainstream sustainable development literature, although clearly it has a huge effect on the debate. All of the schools of thought on the subject start with a value judgement which involves an assumption that a certain model is the best way forward. These assumptions are based on some interpretation of data, but it is seldom a full set of evidence. The bias inherent in the value judgement is seldom addressed, possibly because it is deeply ingrained in the theory, openly political or the experts chose to ignore it. A good example of this is provided in Table 6.2 which shows the effects of the choice of indicators.

The mapping of effects of value judgements is a priority which needs to be properly aired, although it may be the ultimate sacred cow, i.e., untouched and untouchable. Few experts dare to admit that part of their argument is judgmental rather than factual. It may also be that the two factors of *democracy* and *value judgement* are closely intertwined, making their effects difficult to separate out. Most democratic institutes recognise the problem of perception versus fact, and make allowance for the need to split out the value judgement from the factual and incorporate both in an acceptable judgement.

6.2.4 Democracy

Democracy is an important element of any attempt at sustainable development. Colleagues at the Institute have identified this as the key to future debate (Hopwood *et al.* 2000). The ESI checklist approach (Chapter 2) similarly identifies this factor as an important element. It was noted in the last section that democracy and value judgements may be closely intertwined. A value judgement has to be accepted as part of the process within the sustainable development debate because there are so many unknowns, and decisions need to be made on how to include this.

Democracy, on the other hand, points towards who is involved in the decision, an issue that involves principles as well as practical issues. Experience at the Institute has indicated that the widening of the analysis that comes from the study of sustainable development entails acknowledgement of greater shared responsibility – at the global level there is a growing acceptance that more democratic systems of governance result indirectly in greater acceptance of the need to deal carefully with the environment, social and economic issues.

Even at the local level, change management theory, which is discussed in the next chapter, promotes wider participation as best practice because it results in greater embedding of necessary change, an important aspect of sustainable development. Ownership, consensus and the capture of all good ideas relies on participation which, in turn, brings a need or desire to include a wide spectrum of views in some sort of democratic means.

Experts frequently confuse lay people in presentations on the subject. Thus, the expert's role in sustainable development may run counter to democracy. Language can be a significant barrier and the current define–re-define approach has become an easily exploited area for new language to thrive, which in turn makes understanding the basic concept of sustainable development more difficult.

In contrast to the first three factors in this section, the subject of democracy does feature in sustainable development literature, and is the most visible of the factors highlighted in this chapter. Typically, however, it is restricted to one preferred system as an inherent initial assumption. It has been included in this section because it often arrives in the debate as an unchallenged starting point, with little reference to its effect on process and end-goal.

It is not a straightforward subject since many of the arguments would suggest that greater democracy and more people involved in decision-making are the best way forward. However, this itself presents practical problems:

(1)   Decision-making often becomes difficult with greater numbers of people involved since larger numbers bring greater diversity of opinion and full accommodation becomes increasingly difficult, especially where some initial pain is involved before fuller benefit.

(2)   Economists suggest that well-meaning democratic governments tend to establish welfare as a top priority, but this has a detrimental effect on voluntarism and causes loss of social

capital. Views of which form of democracy are best are always likely to excite heated debate with no one clear solution fitting all circumstances. However, despite these difficulties its effects must still be considered.

The linking of democracy to indicators, performance management and evidence has been embraced by governments but with mixed results. In any approach taken to a subject such as sustainable development, where it is acknowledged that a range of complex issues all need consideration, there is a danger that such theory may become unworkable. The UK government now uses 600 indicators to measure its progress against targets. This is a huge number of targets, clearly requiring cross-reference for a full and complete but complicated analysis.

6.2.5 Efficiency

Efficiency takes many forms, but it lies at the heart of many sustainable development problems and needs fuller acknowledgement. The arguments discussed in Chapters 1–4 all need to be efficient to show best practice. In Chapter 5 there is a strong thread of evidence that optimal, efficient solutions are more sustainable than the pretence of a perfect balance between social, economic and environmental arguments. However, there is a clear and important link to a reasonable, acceptable currency or measure of development.

Providing the best solution for the widest population (which may include reference to an environmental interest) or minimising the negative implications of actions will not suit all, nor will it bring a perfect balance, but it will often represent the best way forward. A prime example of this is the debate that surrounds energy efficiency as a key component of a global solution to energy management, climate change, environmental improvement and social improvement. Thus, efficiency in one sector has a knock-on effect across the spectrum of sustainable development interests.

Many of the arguments put forward in the promotion of efficiency have come from business and economics schools. The issue of the currency of measuring efficiency success and whether it is appropriate in this case therefore often arises from the narrow evidence base of economics. However, even within this narrow base it has been noted that there is often confusion between efficiency, a competency in performance, versus effectiveness, producing the intended result. Thus, it is important that all schools within the sustainable development debate need to consider the subject in a consistent manner.

A more interesting example than energy is the production of food. The production of food has seen impressive efficiency gains since 1970, when viewed in terms of production per hectare or production for a given effort. In the period 1970–1989 cereal production increased from 1.2 to 1.8 billion tonnes, livestock in existence increased by 18% and fish catches increased by 67%. Thus, many in government and economic circles saw great success attributed to efficiency gains. However, this has been accompanied by the collapse of fish stocks through overfishing and problems such as mad cow disease brought about partly through intensive farming methods, bringing environmental and social ills. More food has been produced than can be consumed, but still famine and gross surplus occur across parts of the globe, suggesting inefficient distribution through current systems (Middleton *et al.* 1993).

Thus, efficiency purely in terms of production per hectare is a poor measure for this particular subject and many would suggest it is actually a measure of overexploitation beyond natural limits. The arguments show clear problems with language and the currency of success, since both sides are still seeking a similar goal, the most efficient use of the resource.

6.2.6 Space

Space is a particularly important theme in, for example, the specific context of cities, where space is at a premium, reflected in, for example, land values. It is also an important factor at the human level since there are many cultural issues that have developed around the concept of personal space.

There have been some attempts to deal with the subject of space in sustainable development literature through the need to look at carrying capacity of natural systems, bio-productivity, footprints and population density, but it is often indirect rather than direct in nature. Planners provide some interesting debate on the subject (see the SUSPLAN case in Chapter 5 and, again, in Chapter 8), and there appears to be a growing recognition of the connection of this subject to sustainable development in the Netherlands, a country that is densely populated and which sees space as a particularly acute part of the debate.

It is, however, a very difficult concept because space, by itself, tends to have an abstract value rather than a measurable value. It is connected with the rather fuzzy subject of beauty. It does, however, lie within the sustainable development equation as a factor because economically, socially and environmentally, all systems need space to flourish.

Table 6.5 has limited value because the surveys differ in scale. However, the important point to make is that space appears to make little difference to the economic, social and environmental health of nations at present. It may be that this scale is a little crude and that the comparison needs to be more local. However, if the study focused on, for example, public space or population density of inner cities in say, Calcutta, Singapore and the expensive parts of Manhattan, there is a suspicion that, again, there would be no apparent correlation between economic, social and environmental health. There may, however, be a threshold beyond which even wealthy areas suffer.

Table 6.5 Population density versus measures of economic, social and environmental health (1999). [Sources: World Bank 2001a (survey of 207 countries) for population density (rank in brackets) and Gross National Income (GNI); United Nations Development Programme 2001 (survey of 162 countries) for the Human Development Index (HDI); Esty *et al.* 2001 (survey of 122 countries) for the Environmental Sustainability Index (ESI).]

Country	Population density (people/km^2)	GNI/capita 1999 Rank	HDI 1999 Ranking	ESI 2000 Ranking
Singapore	6384 (1)	22	26	65
Bangladesh	981 (2)	170	132	99
Mauritius	579 (3)	81	63	46
South Korea	475 (4)	54	27	95
The Netherlands	466 (5)	16	8	12

Space is a subject that lies at the border of missed or hidden. It is inherent in much of the green debate, but features less in other schools. An important query lies in whether *efficiency* and *space* are inherently linked or do *space* and *population* form a deeper link than the table suggests. Certainly there are arguments for seeing the green argument as promoting the best use of space, where efficiency plays a large part in the definition of 'best'.

For example, issues such as waste minimisation and renew/ recycle have some roots in efficiency and space. A key factor such as *waste minimisation* assumes, through minimisation of all waste, that humanity will reduce cost, space and resource use, freeing them up in an efficient manner for conservation or use elsewhere. The drive to *renew/recycle*, which often accompanies this, often relies on the consumer finding the time and space to store or recycle. Increasingly this becomes the responsibility of the individual consumer to find the space to implement the policy at his or her doorstep. This is possible if there is the space, but much more difficult in a confined living space such as a flat.

Thus, space, a difficult concept, finds its way into many corners of the debate, but often through the other factors of efficiency and population.

6.2.7 Population

Population is an interesting factor in the debate on sustainable development. It again features in many corners of the discussion; through sustainable populations of wildlife, through the effects of overpopulated urban areas and through the effects of population movement across the world (Table 6.6). It often seems obvious that many of the world's environmental ills would be lessened by the existence of less people on earth, and that a debate that included this issue would be useful. The debate, for example, on a return to subsistence living appears to make an assumption that population density would be such that this would cause less rather than more harm.

However, overpopulation appears to be a key untouchable issue for many schools, with the issue remaining on the fringes of the debate for two main reasons:

Table 6.6 Population density versus comfort.

Source	Population density (persons/km^2)	Reasoning behind density definition
Esty *et al.* (2001)	<5	Suggested point at which environmental stress occurs
Mumford (1961)	<25	Space needed for subsistence living (hunting and gathering)
World Bank (2001b)	46	Current world average for population density
Girardet (2000)	4550	London as a benchmark of unsustainable ecofootprints
Sierra Club (Sullivan 2001)	>123,500	Urban density necessary to save hinterland of US from degradation

(1) Historically, population control has been associated with less attractive aspects of government, i.e., fascism, prejudice and pogroms. It has often concentrated on a belief that there are too many poor people rather than too many people overall (Streeten 2001) and concentrated on certain parts of the population.

(2) Acceptance of the problem raises issues on freedom of choice and freedom from state control which appear to be simply too challenging for an open debate, although the issue of population growth has been confronted in some developing nations with birth control programmes, while the developed world has seen a natural slowing of population growth.

Meadows *et al.* (1972) highlighted the issue, but it since seems to have subsided into the shadows of the debate. Streeten (2001), although not writing on the subject of sustainable development, raised the issue of migration control and queried whether global free trade was sustainable if it was not accompanied by global free migration of people. Other factors such as population density (for both man and other species), population movement, sheer numbers and the type of population (poor versus rich etc.) all form part of the debate but are often areas too contentious for clinical cool debate.

It is clear, therefore, that the issue has never been simply one of overpopulation. A more sophisticated version of the argument suggests that (Miller 2000)

Population × technology × consumption = impact.

A reduction of any one of the three factors reduces negative effects. This has the effect of focusing on the developed world's greater impact since its technology and consumption have more effect than poorer parts of the globe. The same reference has, for example, shown that one US child has the same impact as 30–100 children in poor countries. It can thus be shown (Table 6.7) that the population of the USA (275 million × 30 – 8,250 million) has a greater impact than the combined populations of Asia, Africa and Latin America (5,500 million).

Thus, just as with the previous factors, population is a complex subject. There are no easily defined relationships between population and the sustainability of developments. However, on both a global and local scale there is an issue of carrying capacity and the threshold beyond which this becomes unsustainable or causes the problem to be displaced.

Table 6.7 Population density v. ecofootprints (Chambers *et al.* 2000, World Bank 2001a).

Country	Population density (people/km^2)	Ecofootprint (ha/person)	Ecofootprint (% of US)	GNI/capita ranking
Singapore	6384 (1)	6.6	69	22
Bangladesh	981 (2)	0.6	6	170
South Korea	475 (4)	3.7	38	54
The Netherlands	466 (5)	5.6	58	16
USA	30 (169)	9.6	100	8

6.3   Placing the seven factors in a suitable context

Can the seven highlighted issues be factored into a definition of sustainable development and, in particular, the need to consider start-point, process and goal. Table 6.8 shows a summary of some of the evidence of the factors at work and considers whether they are to become part of the wider debate. Greater acceptance within

Table 6.8   Evidence of the seven factors.

Factor	Evidence of their effect within the text	Implications – start, process, end?
Scale	Berwick case study (Section 6.2) Regeneration chaining (Section 6.2)	*End-goal definition only at present?* Needs 'appropriate' scale rather than 'big is good' – 'small is good' extremes
Displacement	Regeneration chaining (Section 6.2) Mango problem (Fig. 5.3)	*Not defined – inclusion at all stages.* Needs universally accepted evaluation tool
Value judgments	Leads to the dogma associated with Chapter 2 v. Chapter 3 arguments	Allows assumptions without evidence. *Accepted at start but has role in process and end-goal stages*
Democracy	ESI case evidence (Table 2.1) and implicit in attitude to decision-making	*Not defined but included at all stages.* However, the role of players throughout the process is not clear
Efficiency	Waste minimisation (Section 6.2) Energy efficiency (Section 6.2)	*Not defined.* Needs inclusion in debate and to be broadened beyond economic definition
Space	Development of protected or conservation areas (Table 6.6)	*Not defined.* Needs inclusion in debate and further conceptualisation
Population	Pressures of migration (Section 6.2) Pressure on water resources (Section 3.1.2)	*Not defined.* Needs inclusion in debate and desensitisation of debate

the debate will call for lateral thinking and a breaking down of the barriers developed by all schools – barriers of language and barriers of the traditions of taboo – and some of these issues are addressed again in Chapters 7 and 8. Table 6.8 shows the immense difficulty of pinpointing the effects, although there is growing recognition of their existence.

Figure 6.2 shows the myriad relationships between the seven factors. With such a complex web it is easy to see how the factors can be missed, forgotten or how analysis is unable to conclude their true effects. At this point in time, therefore, it is difficult to judge how these factors fit into our original framework question of starting-point, process or end-goal. It must be pointed out that the combination of these seven factors does not represent any attempt at a complete theory of sustainable development. Indeed, it has been viewed only as a starting position. There may be other gaps in coverage which have even greater effect and become obvious through further discussion.

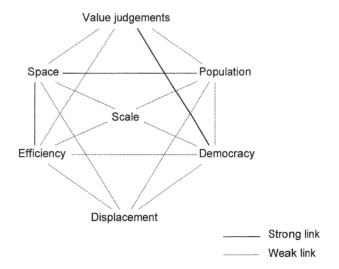

Fig. 6.2 The seven point diagram, showing linkages between factors.

Two recent conferences on specific issues associated with sustainable development (SUSPLAN 2001, Sustainable Information Society – Values and Everyday Life Conference 2001) have proved a good ground for testing the existence of the seven factors. References from the two conferences (again in Table 6.6) suggest that the effect of all seven factors is now actively being considered, though often hidden under different terminology. An important conclusion from the conferences was the confirmation that evidence bases are viewed as the critical next step and, crucially, human behaviour, the subject of the next chapter, is a key factor.

6.4 Summary

It was noted in Chapter 1 that the lack of an agreed, unifying end-goal or clear mechanisms to assist standardising sustainable development processes are a root cause of the current piece-meal approach to sustainable development. Chapter 5 returns to this theme and highlights the problems of a universal process map. This chapter has highlighted some of the gaps in the current debate which stand in the way of further progress towards a process map. This appears to be the result of oversight, bias, legacy and misunderstanding, and suggests that an individual's perceptions, language and thinking are key issues. The individual's response to key concepts within sustainable development will therefore be studied in the next chapter.

The evidence of the importance of the seven factors is substantial. Identification is the critical first stage, and beyond that the effects of these factors can be noted in the current debate, although they are often hidden behind rhetoric, false information and other factors. The effects of democracy, efficiency, space and population are generally acknowledged, but their full effects are hidden and not fully analysed as a result. The first three factors, scale, displacement and value judgements, need concentrated analysis if they are to be seriously considered in the debate. Highlighting the seven factors as omissions in the current debate, there is a risk that (1) they will be added to the already long list of priorities, which is inclusive but unmanageable, and (2) these new factors will become the focus of debate to the unhealthy exclusion of other equally important factors.

The question remains, therefore, whether a useful theory can be developed without further unnecessary complexity, an issue that will be addressed again in the next chapter.

7 Breaking Down Entrenched Positions

The previous chapters present sustainable development as a complex cross-disciplinary subject. There are suggestions that the subject involves opening up thinking and approaches beyond the traditional single-discipline type approach so that solutions, which look good from one angle, can be fully assessed from all angles. Many would argue that this already occurs, and where it does not occur it is because it overcomplicates matters, making decision-making more difficult with no obvious improvement in result.

There are two areas of study that remain to be discussed:

(1) How projects, developments and progress are conceived or planned, and whether this can improved. Two key aspects of this are the manner of dealing with *cross-disciplinary* factors and attention to *futurity* as an issue.

(2) The concept of *change* and human reaction to it. Much of the debate points to change, whether it is the incremental change of the status quo advocates or the radical change proposed by the transformationists. Change is a difficult concept for humans and has spawned a whole new subject area of management theory (Carnall 1995).

Do the mistakes of the past, highlighted by the environmental and social sectors, warrant wholesale change and the study of development from a cross-disciplinary approach rather than combining single-discipline expertise? With better tools for conceiving, planning and changing, could humanity avoid the mistakes of the past, build futurity into projects and develop a language and framework for a full spectrum of players to use? Can future decision-making be better informed, more democratic and can more complexity be avoided? This is a tall order given the obvious queries on the current evidence base, the risks of prediction and the political dogma often attached to discussion on sustainable development.

This chapter enters the debate by looking at the three basic subjects of cross-disciplinary study, the 'tools for future decision-making' and the approaches to change management. In each case the challenges are similar: what tools are available, what constraints are introduced, how complex are the methods and where does the individual fit into the picture? Much of this revolves around the individual's responses to the systems.

7.1 Cross-disciplinary study

The International Centre for Integrative Studies (ICIS) is an institute set up in the Netherlands specifically to research the subject of cross-disciplinary study and integrated assessment techniques. ICIS has mapped how integrated land management from the Egyptians through to rotation of crops in the Middle Ages shows that integration of agricultural knowledge, irrigation and weather forecasting has been practised since ancient times (ICIS 1999). It is therefore not a new subject, although it appears to have gone through a rebirth in parallel with the debate on sustainable development. Integrative studies in the modern era appear to have roots in the early 1970s with the dawn of computer simulation models and techniques. Many of the early studies were linked to unravelling the secrets of the environment in its broadest sense. One interesting participant in this has been the insurance sector, which has availed itself of the new techniques with more complex risk assessment, and new markets have developed in areas once thought of as uninsurable. Thus, a link is seen between better risk assessment and better analysis, which may lead directly to more sustainable development.

While the subject draws heavily on science and technological progress as its foundation it is acknowledged that techniques and knowledge will always be incomplete and that answers are not universal. Thus, it is accepted by this school that the subject is not fully objective, and there will always be an element of political choice in decisions or solutions (ICIS 1999).

The term cross-disciplinary means different things to different people. It is often confused with the terms interdisciplinary or multidisciplinary, or other substitutes such as integrated study or multi-modal study, etc. Thus, language and the meaning of words are critical where work crosses over a number of boundaries. Early experiment in our own Institute (Mawhinney 2000) showed that social scientists and engineers have a different perspective of even the most basic building blocks of sustainable development. A word such as 'theory' can, through reference to a dictionary, reveal very different meanings which translate into different approaches and, in turn, into different end objectives. Taking the word 'theory' as a starting point and using the Collins dictionary (1998) there are a number of different definitions which include:

- Theory (1) – a system of rules, procedures and assumptions used to produce a result.
- Theory (2) – an ideal or hypothetical situation.

Within the group of researchers at the Institute there were social scientists, engineers and other specialists, some of whom identified more with the first definition while others identified with the second. For example, the social scientists formed a view that accepted an element of political bias within a subject as the theory (definition 2) and then tested it, while engineers observed and mapped a system or approach, with little recognition of any possible initial political bias (definition 1). Thus, fundamental differences in language need to be addressed before cross-disciplinary study can be achieved.

The rationale for dedication of specific resources to the study of complexity and complicated cross-disciplinary issues rests on two beliefs:

(1) The assumption that modern society has become so complex that integration plays an important role in life; this is the result of the increasing complexity of technology, and the increasing political integration and economic integration that accompany today's trading and living patterns.
(2) An acceptance that there is a gap in the thinking and logic of the disciplines; it is of crucial importance that bridges be built to overcome the different gaps. The speed of information movement can only make this more complex.

Integrative studies are particularly required in situations involving 'causally linking processes' which have different time-scales, spatial scales and dynamics and may thus be measured in different manners. A good example of this is the climate change debate from Chapter 3, where the fundamental factors would include those shown in Table 7.1. Thus, the dimensions of the problem are a complex web of factors working to different time-scales in different geographical scales, but each part is too important to dismiss.

However, major issues such as climate change are already being addressed by professionals, and reference to the internet reveals many who already see themselves as cross-disciplinary. Geographers, biochemists and architects would all claim this

Table 7.1 The different aspects of climate change.

Factor	Time-scale (years)	Spatial scale	Example of effect
Human interventions	1–70	Can be very local	The worst coal-burning power stations
Socio-economic processes	3–5-year plans	National	Budgeting for disaster planning
Land cover processes	20–30?	Continental	Desertification
Atmospheric/climate processes	100+	Global or intercontinental	Global warming
Ecological impacts	100–1000	Geographical (e.g., watershed)	Shifts in patterns of rainfall

status. General managers in business would also argue their need to integrate, to see all sides and be cross-disciplinary. Many would therefore query why the subject should be set aside as a special new interest when it has a history and is an inherent part of everyday work.

Interestingly, ICIS (1999) suggests that gaps are formed in major studies on the basis of differences of processes and tools. Our own experience at the Institute suggests that the problems are deeper and differences emerge in language, culture and evidence priorities, as noted earlier.

Identifying barriers

At the Institute an initial 18 months was spent trying to brain-storm the subject of sustainability and sustainable development. A forum was created within the confines of the Institute to study the issue of a common definition for sustainable development, particularly in the context of the city. This inevitably led into the theory rather than practice-led approaches to the subject, and a weekly brain-storming session with six academics was arranged. The team represented a balance of subject disciplines across social, economic and environmental, although initial sessions were hindered by three problems:

(1) The language barriers: language is undoubtedly different and, as pointed out earlier, even words such as 'theory' have different connotations to the different disciplines. At the Institute the method used to reduce this problem was to go back to first principles and redefine in simple terms.

(2) The expert barrier: it was interesting to note that when discussion strayed too far from the grey mid-ground between the disciplines and into one of the disciplines, invisible but perceptible barriers arose as expertise felt the need to defend itself, a natural human reaction. Again, a simple solution was sought by breaking the group into smaller, more manageable groups.

(3) The evidence-base: perhaps in line with language differences is the degree of comfort that disciplines have with various forms of evidence. It is obvious, for example, that engineers rely heavily on arithmetical solutions, social scientists, by and large, preferred qualitative information and arts often look to illustrative evidence. While it can be argued that good researchers should communicate and use all three it is clear that preferences play a strong part in their approaches.

Cross-cultural problems with communication are a common theme in many books on the issue of workplace interaction (Berry and Houston 1993), although they are typically associated with cross-national differences rather than cross-discipline. Identifying differences in language, the purpose of communication and dealing with distortions of communication channels at the organisation level, and then identifying how individuals or groups interact in a particular situation are viewed as important components in deciphering the problems.

Cross-disciplinary study can potentially be an expensive exercise requiring a range of experts to sit together at the same time and work together on issues at hand. It is therefore important to correctly identify whether the complications of study subjects warrant the additional resources to study them fully, assess what is the demand for integrated types of studies and search for simpler alternatives.

The traditional approach has been *reductionism*, where a complex subject is split into a number of manageable elements and each element is assessed individually. Special teams can be employed to study complex parts of the subject in a holistic manner. This approach forms the basis of much of mankind's approach to life. Examples would include the division of government into ministries of health, foreign affairs, home affairs, transport, etc., or universities with their faculties of art, social sciences, engineering, business, etc. Special project teams are then formed to study difficult subjects such as disaster response or major change programmes, restricting the cost of cross-disciplinary work. There are big advantages to this approach; it is often sufficient to believe that the whole is the exact sum of the parts and it is certainly easier to conceptualise and properly analyse issues in discrete bits.

By contrast, the alternative *holistic* approach assumes that each element has a cause and effect which, in turn, has side-effects. Many of these side-effects occur beyond the realms of the element's boundaries. Thus, the whole is greater than the sum of the parts because of the interaction across the parts (Fig. 7.1). This therefore leads to the need for integrated assessment to map the extra effects. Two examples of this would be climate change (Table 7.1) and Girardet's mango problem (Fig. 5.3), both of which need a wider analysis than that gained by breaking them into their elements.

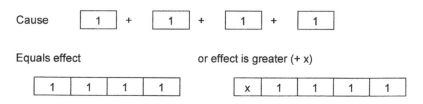

Fig. 7.1   Matching cause to the full effects.

ICIS (1999) suggests two reasons for the pursuit of holistic approaches rather than the cheaper, easier reductionist approach: (1) the missing of important cause and effect chains and an acknowledgement that scientific 'fact' is often incomplete; and (2) the increasing complexity of key factors supporting lifestyle and work, such as policy instruments or technology leading to greater knowledge and shorter life-cycles for planning, etc. It is also suggested that fuller risk analyses can be completed as a result of this type of approach, although at greater cost.

This has clear implications for the study of sustainable development, with its myriad of elements and its emphasis on inclusiveness. It has the potential benefit of ensuring that a particular problem is placed in a proper context, i.e., it is placed in the broader context of other issues and possible responses. However, it is acknowledged that integrated approaches must be appropriate in terms of time-scale, spatial scale and rules on equilibrium. ICIS (1999) further suggests that some simple tests are needed to check whether an integrated type of assessment is necessary:

- Issues with a short time-scale or a high degree of certainty can be eliminated.
- Flow-diagram approaches to the systems involved can indicate potential interactions which will complicate a reductionist approach.
- Systematic assessment of the drivers, effects and implications can indicate complications which need analysis.
- Significant differences in the relevant time and space scales involved in the subsystems which make up the system can indicate added effects.
- An assessment of whether the problem is 'complex' (with many interrelations between processes and subsystems leading to added cause and effect) or is the problem simply 'complicated' by the number of processes?

Subject matter that can be eliminated or simplified through these tests can then be approached using reductionist methods. Where this is not possible an integrated assessment may be necessary (Table 7.2).

Table 7.2 A summary of reductionism versus integrative studies.

Information handing stage	Reductionism	Holism
Input	Information is produced and digested in series	Information is produced and digested in parallel
Processes	Traditional, well tried	Requires new methodologies (see Table 6.3)
Output	Series of outputs brought together	Often whole picture scenario based
Advantages	Simple, traditional, works for many situations; human scaled – easy to understand	Fuller analysis
Disadvantages	Misses important links?	Complex, methods still not fully tested; no guarantee of accuracy
Costs	The sum of the parts	Additional resources required

The arguments that are put forward in favour of integrated study approaches are an interesting confirmation of the previous chapter. The problem of scale is strongly highlighted. The suggestion that scientific fact is incomplete implies that value judgement and political decision creep into all matters. The problem of finding hidden effects beyond the boundaries of each separated element implies displacement. The need to check whether a problem is complex enough to warrant an integrated study suggests the importance of efficiency. Many basic system problems are dynamic rather than static in behaviour, and many have cross-scalar and iterative behaviour affecting the system, causing extra effects to those predicted by reductionist approaches.

However, a number of issues remain to be addressed before there is any fuller acceptance of the widespread use of integrated studies.

- At the human level there are the problems of coping with the complexity required in attempting to cover all of the gaps. Integrative approaches are infinitely more complicated to think through than reductionism, and an overreliance on computers quickly exposes analysis to mistakes.
- Uncertainty is not eliminated. It is merely addressed in a different fashion to traditional methods. A range of methodologies for integrated assessment are available, some of which will be studied in the section on futurity tools. They break down to (1) obtaining more information and using computers to analyse the information, presumably because the human brain alone is insufficient, or (2) involving more participants in the information gathering to improve ownership and spread of information.
- The question therefore arises of what are the additional benefits from the increased complexity of preparation, analysis and implementation? Is there a supply of good information sufficient for integrated studies? Is there demand for the increased accuracy that it promises?
- It needs bench-marks of what simplification is acceptable and how to address differences in levels of knowledge for particular subsystems. Examples given in much of the literature are obviously at either extreme – what about the grey areas?

The difficulty of dealing with complexity to the n degree and the added cost lead to a tendency to replace integrated study with a reductionist approach. This is difficult to counter when there is no guarantee that greater detail will produce more accuracy. Only further examination of the subject and success with the methods will convince people of the need for change.

Can reductionist methods be improved to take account of extra effects? An easy method which may help relies on the Pareto principle, which highlights the effects of the vital few and the trivial many (Tangram 2001). The principle suggests that there are many examples of the 80:20 rule where a simple analysis provides 80% accuracy which is sufficient for most purposes, and avoids all the problems of more but unneeded accuracy and an overcomplex analysis. To benefit from this, however, the analysis needs the setting of a clear end-goal, an observant eye to detect the possible short-cuts and clear measures of success.

7.2 'Change' as a concept

Restructuring, globalisation and IT made significant impacts on business in the 1990s, provoking a surge of interest in the study of change and its effect on organisations and people. The objective was straightforward: change was recognised as necessary and it was becoming increasingly clear that many businesses were doing it badly. Palmer (1998) points out that rapid change is not a new phenomenon. Business in the developed world has, for example, come through fairly rapid major transformations such as the move from an agrarian society, the industrial revolution and mass production before the current period of change.

A management school theory of business transformation and change management may not appear to be an ideal starting point for sustainable development, but it still represents a useful introduction. Arguments to date would suggest that the problems in defining sustainable development are as much management and human behavioural as based in the subject itself.

Much of the literature on change management concentrates on how to better manage inevitable change, an issue that all sides would agree is pertinent to sustainable development. Business transformation for a company occurs when the organisation, its resources and people change direction. This can result from a strategy-driven approach, where the change is planned, or from reacting to an unplanned problem or opportunity. It can be implemented on the basis of business processes, compctencies or

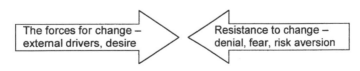

The forces for change – external drivers, desire ▷ ◁ Resistance to change – denial, fear, risk aversion

Fig. 7.2 The change force field.

creativity/learning, i.e., using the strengths of the organisation as a whole, its people or learning new skills.

It is now widely agreed (Carnall 1995) that effective change requires (1) organisation, (2) a desire to change and (3) clear objectives or goals. Evidence of both top-down and bottom-up achievements in the change process is vital, in order to ensure consistency, ownership and integration throughout the organisation. This is because change involves adjustments to a set of dilemmas, attitudes and behaviours, all within the context of unpredictability, and all staff need to feel involved in the ownership of the problem and solution.

Thus, it is a complex operation since both organisation and personnel need time to adjust, and change management programmes frequently go wrong. It has been stated that 75% of all transformation projects in business fail (Bullet Point 2001). Much depends on individual or human response to the change and the nature of the implementation of the transformation. Carnall (1995) suggests that humans react in a five-stage manner to major change (Fig. 7.3).

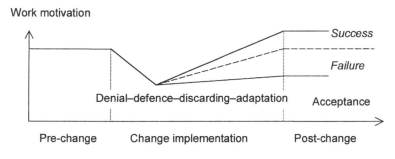

Fig. 7.3 Human response to change (after Carnall 1995).

The five stages of denial, defence, discarding, adaptation and internalisation can be helped or hindered by the speed and nature of implementation of the change. In the first stage the need for change is denied and resistance results. In the second stage the reality of the need for change becomes dominant, but a self-defence culture still prevails. By the third stage people begin to look forward and discard the belief that previous methods were best. In the fourth stage individuals start to adapt to the new or proposed systems and the fifth stage sees staff feeling ownership of the new systems or ways of working. Obviously this is an idealistic model and at each stage there will be people left behind or who do not move into the next stage. However, it is claimed that the model represents the majority of people in a well-managed, changing organisation (Carnall 1995).

Palmer (1998) believes that collaborative change rather than coercion creates the more profound and long-lasting change, since it ensures ownership and leads to greater acceptance of responsibility throughout the organisation (Table 7.3). However, there are many examples of good change resulting from forced or coerced change, and the common lesson appears to be preparation and communication rather than simply ownership.

Table 7.3   The management of change (after Palmer 1998).

	Simple change	Transformation
Collaborate	Participative evolution	Lead by persuasion
Coerce	Forced evolution	Dictate

A critical issue is how many people need to be involved in the preparation and communication of the change. Stakeholder theory suggests that a wide range of people are affected by changes that often seem internal to one organisation – suppliers, clients, policy-makers, etc. – and a debate rages as to how many have rights or an interest in any major change.

An interesting approach to the subject of change management comes from Senge (1996). He argues that change is inevitable and must be embraced within successful organisations, that corporate leaders seldom have the power to force change, but their main role is to provide clarity for the organisation, and that significant organisational change requires local team leadership and a change at individual response level.

He notes that, on the one hand, the precondition of team-building is to have individuals who believe that they need each other and that, on the other hand, different parts of a corporation work to different time-scales, thus creating great complexity. Problems in organisations arise from issues such as the wrong solutions to problems in the past, good-intentioned intervention causing problems elsewhere, short-term benefits being easier to measure than long-term profit, cause and effect having different time-spans and optimal solutions not being the fastest or cheapest. All of this would suggest many parallels to the lessons coming out of the previous chapters.

His solution to the dilemma is to develop businesses as change-orientated organisations, and hinges on the belief that the learning involved in change management needs five disciplines:

(1) Systems thinking or an acknowledgement that the elements of a system continually affect one another and that feedback loops are critical to the system (interestingly his feedback loops suggest the presence of displacement);
(2) That all individuals must understand the learning process preparing the individual to be a part of the team;
(3) That models must include 'images, assumptions and stories' just as suggested in the previous section;
(4) That a key requirement is a shared vision and how it fits into the larger world;
(5) That all of this is fed by team learning (as opposed to team building), a process of open learning rather than closing minds around a belief, direction or concept.

Senge promotes two methods of learning: (1) catalytic, where new experiences are deliberately manufactured to develop creativity, and (2) proper learning for individuals through everyday experience and their attitude to this experience. Both of these methods allow the organisation to capture the best thinking for different teams and time-scales.

While most change management theory concentrates on the corporate world, it is believed that three socio-cultural factors are critical to change in general for all organisations; learning styles, attitudes to risk and risk-taking and motivational drivers.

Many people learn or acquire new knowledge and skills through four main routes (Carnall 1995): the *activist* shows enthusiasm towards trying anything new. They learn through participation and may get bored after one or two experiences. The *reflector* observes and analyses as their learning process, marked with thoroughness and caution. The *theorist* needs to include observations into a grander theory before accepting new learning, looking for coherence across a system. *Pragmatists* actively look for new ideas to test and quickly assimilate, being practical people. Often the different individual learning styles are best accommodated by different parts of the change cycle, i.e., it may be possible that individual learning and the process of organisational change can be matched in choosing the players and styles for major programmes of change.

'*Fear of failure*' is recognised as a major issue for many

organisations, often manifesting itself as risk-aversion. The prospect of change or transformation is recognised as a particularly uncertain period likely to increase the 'fear of failure', often at the denial and defence stage. Consequently, resistance to change is an important but difficult constant in many change programmes. It is therefore important in change programmes to address the issue of attitude to risk-taking.

Maslow has produced a theory which has been used by many to interpret attitudes to work and change at work (Royal Navy 1995). It postulates that individuals have a hierarchy of needs; at the lowest level are the physiological needs, next comes safety, social needs, esteem needs and finally self-actualisation needs. In other words, people make sure they are physiologically satisfied with food and shelter, etc., before they consider safety.

At the next stage, once both of these are satisfied, then people will consider their social needs, and whether they are held in high regard by others. The pinnacle of achievement is the belief that the individual has fully achieved his or her potential in his/her own eyes (Royal Navy 1995). It is believed that individuals are constantly motivated to move up this pyramid to the point of self-actualisation at the top, which is itself a self-motivating condition (Berry and Houston 1993).

How does the above discussion on change management help to explain or deliver sustainable development? First, there is the belief that, as with complexity and integrated analysis, much is dependent on the individual and his or her response to a situation or set of circumstances. It is therefore helpful to understand how individuals respond to complex issues such as sustainable development.

Second, the factors raised in the previous chapter are shown to be general problems associated with all change rather than specific to just sustainable development: the need for wide ownership of problems and solutions, the consideration of democracy, the effects of individuals' value judgements and motivations, the identification of displaced problems and scale differences by Senge and the need for shared sets of objectives.

Senge's theories in particular are helpful in confirming what the problems are, but it still leaves us short of many answers. Beyond the above conclusions, however, there is only speculation. Within learning styles there is a possible explanation for the various schools of thought; the pragmatists look to test out new ideas in the field (Chapter 5), the activists look for new ideas, the theorists (particularly economists) are unwilling to accept new learning without a grand theory and, since no one route has found the best way forward, all three continue to promulgate their own theories.

Again, with sustainable development there is a universal acceptance that change is required at the global level and that a change programme is required. Fear undoubtedly continues to play a part in the level of acceptance of the need for change. In some areas, such as waste and climate change, where the evidence base is available, there is still strong resistance to action on the subject.

Figure 7.4 is an attempt to describe the various schools and how they fit with Maslow's theory. The different directions of the schools, if true, actually represent demotivational forces for the opposing views, and this helps to explain the deep divide between the two schools, and the lethargy in accepting the views of either extreme. Thus, change management contributes to the debate on sustainable development in explanation of contributory factors. It suggests that humanity is not well enough organised at this point to deal with a complicated subject such as sustainable development. It is, however, ultimately disappointing in providing solutions. The success or otherwise of proposed change is dependent on many factors, and it clearly is a difficult subject psychologically for humanity. No solution is easily derived from an understanding that sustainable development is complicated and involves change, both of which much of humanity avoids if possible.

A key factor appears to be the influence of leadership and developing a shared vision of what change and what complexity is

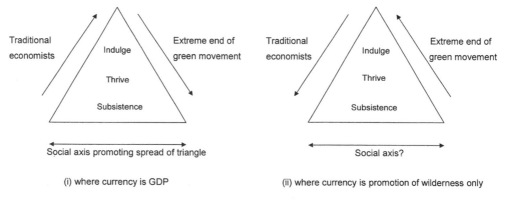

Fig. 7.4   Maslow's hierarchy – Change in what direction to achieve sustainable development?

involved in the decision-making process. This leads into the next section, which has a heavy emphasis on the 'shared vision' element of the formula for successful change.

7.3 Futurity tools in current use

There are many situations where a future event is fairly predictable in scope and a few experts would claim to have mastered the art of forecasting on the back of experience with predictable events. However, the more generalist and the more honest see the prediction of the future as a combination of science and art, with many risks inherent in the exercise.

Futurity tools have become increasing popular as computers and their ability to analyse greater amounts of data have led to the belief that better prediction and planning ahead will result. Since sustainable development involves a large element of looking ahead it is useful to review the tools available, and to study whether they incorporate or make allowance for cross-disciplinary input and change.

Planning ahead is a necessity. The most basic car journey from point A to point B needs a plan of the best route. It will need to take account of the mode of transport, the possibility of other traffic and other factors that may make the journey longer in distance or time, or which may cause other discomfort. In the same way, people plan their next career move, companies produce business plans and public sector organisations develop forecasts of future needs. In all cases there is a need to take account of the best route available and the external factors that may impact on choosing that best route. The big questions therefore are (1) how to assess the best route and (2) the constraints on that choice.

The need to avoid or reduce risk is a key part in the desire to seek a degree of certainty in future events. The management school advocates of strategy development, however, suggest that another primary function of this type of activity is the development of innovation or creativity, together with the increasing belief that change requires development of ownership of the vision/ strategy/future predicted, as noted in the last section. Thus, establishing *certainty*, improving *creativity* and *ownership* are viewed as the cornerstones of good futurity exercises. All three, however, introduce difficulties, made worse since sustainable development as a subject falls within the more complex end of prediction, forecasting and planning.

The simplest of tools will seldom provide sufficient certainty, creativity or ownership for complex subjects, and more complex tools are often required. A number of these tools are increasingly associated with the study of sustainable development, and these will form the basis of this section. However, even the more complex tools can never fully predict the future.

Where there is an acknowledgement that prediction will be difficult, two paths are often available. The objectives are not necessarily the same:

(1) *Getting consensus* – the development of a consensus either across experts or across a wider group of individuals. Where the group comprises experts the intention is often to get the best of expert opinion. Where it is a wider group the aim is often more towards establishing ownership.

(2) *Getting it almost right* – the use of whatever tools are available to eliminate as much error as is possible. This is, however, seldom likely to involve complete elimination of error and risk.

Two extremely important factors in forecasting are the spatial scale and the time-scale. Both these factors are often defined arbitrarily when, in fact, it is important to define the correct boundaries of the forecast exercise. The physical scale of the analysis needs to encompass the environment which includes all major influences impacting on the proposed project or development. The time-scale chosen needs to reflect an appropriate period for the development and profit from the project/development/progress without the need for major change. Both of these factors reflect the need to avoid displacement on the one hand and unnecessary complexity or analysis on the other, as Section 7.1 indicated.

Bringing all of the above into a coherent methodology is a challenge. On the one hand, there is the need for a degree of certainty, some creativity and ownership. On the other, the need to acknowledge, with some accuracy and without added complexity, the parameters that define the scope of the problem and agreed discrete points in the future where progress can be measured.

There are three types of methodology (ICIS 1999), although they clearly overlap in detail: extrapolation, modelling and leave it to the experts.

Extrapolation forecasting is concerned with the prediction of trends, turning points and discrete events, and there are some subjects where this is easier to predict than others. An obvious example is the number of 30-year-old people 25 years from now. Since these people are already born only small adjustments are necessary to account for a small number of deaths over the period. Other subjects are more difficult to predict, such as the world's notoriously irregular weather pattern or many aspects of human behaviour, and these are unlikely to be suitable for extrapolation exercises.

Extrapolation typically involves evaluating past performance, identifying a pattern and extrapolating this pattern into the future in a simple, straightforward manner. The method often relies on reliable quantitative evidence as the basis for extrapolation, together with an assumption that past behaviour will continue in a similar manner in the future. Key examples include Lomberg's assumption of constant energy use in Chapter 3, or the predictions on population increases from Meadows. The advantage is that it is simple and easy, and even lay people can use this type of methology. The disadvantage is that its simplicity may not be sufficient for many events or patterns of behaviour.

Modelling from the simple to the complex Modelling is the favourite tool of the research community and covers a variety of simple and complex tools. The defining difference between modelling and other forms of methodology is the emphasis on a structured process to the approach rather than the outcome, which is the important element for the other two.

At the simple end would be the triangle models developed in Chapters 1 and 5. At the other extreme would be world climatological models noted in Table 7.1. The increasing use of computer simulation as part of modelling is allowing researchers to include more variables, more interrelationships and more sophistication. This added complexity brings new insights, but it also masks the fact that uncertainty has not been eliminated, and it often hides a bias, arising from the objectives of either optimisation or evaluation (ICIS 1999).

Modelling has many of the same strengths and weaknesses of the other two methodologies, i.e., it works very well in some circumstances, but not in others.

Leave it to the experts This is not strictly different from the other two types, since experts will frequently use simple extrapolation or other forms of modelling to produce a prediction.

It is a popular method to appoint 'experts' to produce an opinion or set of options. The result can be quantitative, qualitative or a combination of both. It can also be structured or unstructured, depending on the outcome desired.

The advantage is that it is an easy process, and one where responsibility is assigned to the 'right' people. The disadvantage is that this does not ensure correctness since experts often bring their own bias and the 'experts' may not simplify the message sufficiently for communication purposes or for the underlying assumptions to be understood and properly challenged.

Thus none of the basic methods is ideal. They introduce a degree of certainty within some situations, but they seldom fully address ownership or creativity where needed. They still need a political decision, whether it be the expert researcher's concluding view or that of a politician, on what parameters to include and what to prioritise. The accuracy of forecasting or planning output, as with all information management, relies as much on the quality of input as the method of analysis. Since there is no perfect methodology nor perfect data there needs to be allowances and safeguards to ensure that the information is used appropriately or is not accorded a value that it does not deserve.

Improvement in modelling and computer simulation will continue, and there have been some attempts (ICIS 1999) to produce models that are hybrid and more integrated. There have also been attempts to clarify modelling in terms of complex (few processes each with potentially large effects) rather than complicated (a variety of processes but each with limited effect).

7.3.1   Improving input and output

The changes to predictive methodologies most closely associated with sustainable development work therefore tend to look at the input and output ends of forecasting events or projects. They concentrate on the aspects of the process that are particularly in need of improvement. All, however, introduce new bias, and may actually displace the original problems or limitations into new directions. The list of such methods below is incomplete, as all such lists tend to be, because it concentrates on first-hand experience within the Institute.

Refinement of output
  Consensus
  Indicators
  Risk assessment

Refinement of input
  Scenario planning
  Delphi technique
  Participatory appraisal techniques

The output refinement techniques, through involvement of wider participation or through using the proxy of indicators, have already been discussed. An example is included below.

Leave it to the experts – an output approach

A straightforward approach from the UK has been the development of Foresight panels. These are panels of experts brought together to develop their ideas about the future in relation to specific topics. Examples include crime prevention, ageing population and healthcare. The information from expert discussion is then available to the public or to businesses, from which they can draw their own conclusions.

The advantage of this method is the pooling of expert resource, but, again, it brings no guarantee of success nor can it claim to avoid the pitfalls suggested in the development of the Delphi technique where the loudest expert has the biggest say. More importantly, there is some confusion about whether the main objective is to provide business and the public with the information to meet change or whether it is a tool to stimulate innovation.

The rest of the section concentrates on the refinement of input.

Scenario planning is an easily defined concept that has great potential but is often badly implemented in practice. Scenarios are descriptions of the future. At their simplest, they are a fantasy or aspiration. However, to become powerful planning tools they need to incorporate careful assessment of present and future situations. They need to become a vehicle to broaden perspectives, raise questions and challenge conventional thinking so that they address risks and possible discontinuities in any trends predicted.

Initially a group of people with an interest in a particular project or event are gathered together and probed about every aspect of the project. Experts are brought in to challenge assumptions and to explore issues on the periphery of thinking. All the material is then collated and used to develop a number of scenarios, and from this a strategy is devised which can allow the group to detect change which may force the project in a particular direction and actions to alleviate any risks that result (Eglin 2001).

Scenarios have to be firmly rooted in the present, and political dogma creates bias which renders the exercise useless (Kassler 1994). As with all methodologies there is no one set manner in which the exercise is conducted. In some cases, 'experts' are used for all stages, in others they are used selectively to sift and analyse data which come from a lay audience, while in other cases 'experts' are excluded.

A number of outcomes are possible:

(1) *Exploratory options*, based on a sequence of assumptions moving forwards
(2) *Anticipatory outcomes*, based on an assumed final state and tested by reference back to current trends and situations
(3) *Descriptive-normative scenarios* where the emphasis is on description with and without desirable goals.

They can also be single issue, focused or global in scale, giving a wide range of options. Frequently, however, the choice of outcome is not clear at the outset and this results in a muddling of the goal, process or the involvement of the various players.

The need for reference to a firm rooting in the present can lead to many difficulties for people trying to conceptualise a future beyond a few years. Bias is a constant concern, along with inconsistency in assumptions, lack of transparency from expert input and the tendency of groups to translate the need for a range of scenarios as being best case, worst case and one in the middle (ICIS 1999).

On the positive side, however, constraints and dilemmas can be explored in an open creative forum if all goes to plan in the exercise. A wider set of data can make their way into the analysis. The invitation of wide groups of participants can help to establish broad ownership of future plans. It is further claimed that the method is particularly strong in subject trends with difficult or unpredictable events causing discontinuities or changing priorities.

Glenn (2001) argues that the benefits of scenario planning are twofold: the benefits associated with a group of individuals participating in a planning event through increased awareness, team-building and ownership, and the individual's development of anticipatory skills associated with testing assumptions and questioning possible outcomes.

Table 7.4 presents a sample of examples of scenario planning. While the tool is an interesting method which attracts much attention, its one overriding problem remains the tendency of advocates to fall into the trap of believing it to be predictive.

The *Delphi technique* attempts to address some of the biases that are associated with the human judgement angle within forecasting. Its three characteristics of anonymity, statistical analysis and feedback of reasoning, together with an expert team approach are claimed to present a better balance for tackling the difficult issues with creative approaches.

A group of experts are gathered together to provide the input for a forecast or decision. They are not told each other's identity. The

Table 7.4 A sample of leading examples of scenario plans.

Source	Method	Objectives	Participants
Shell (Kassler 1994)	Experts develop stories Story-telling workshops 3 scenarios	Develop business strategy around energy use and uncertainty	Shell employees through the workshops
British Airways (Moyer 1996)	Team develop process 28 workshops 2 scenarios	Develop business strategy on global airline industry trends	Initially top management but subsequently 280 employees through workshops
North-west England (Ravetz 2000)	Story-telling and research Extrapolation 10 scenarios	Ten-year regional strategy aspiration v. reality assessment	Local experts and decision-makers supported by facilitators
United Nations (Glenn 2001)	11 global nodes collect and update annually 300 scenarios	UN-led think-tank purposes	One thousand experts across the 11 nodes
European Union (van Asselt *et al.* 1998)	Discussions amongst staff development team	Policy exercise called Vision 2020	Internal exercise
Quest (Sustainable Development Research Institute 2001)	Game-type approach, action and consequence	Awareness-raising of ecosystem limits in British Columbia, Canada	Combine eco-experts with interested public and decision-makers

anonymity ensures that the fashionable (as opposed to the correct) do not receive undue weight in the process, a common problem in forecasting. In the process each individual is asked for a forecast or is given an open-ended questionnaire. The responses are then correlated. Those forecasts or views that fall outside the consensus are then asked for explanation or eliminated. The process is repeated until a consensus emerges.

Again, this approach provides no guarantee of accuracy, but it does remove some of the bias. Opponents of the technique argue that the anonymity within the system leaves it open to abuse. The choice of which experts to include does, of course, include a judgement. The central analysis necessary to establish consensus provides another source of possible bias.

A subject such as sustainable development lends itself to the Delphi technique since it is a very subjective subject open to bias and judgement. A good example is the Bremen Partnership Award, which judges the sustainability of urban development and was set up using the technique (Mayank 2000).

Participatory appraisal techniques concentrate on the ownership issue, and often involve approaches that are dedicated to seeking the opinion of lay people only. The techniques used include focus group methods, dialogue methods or simulation exercises which all concentrate on obtaining information from the lay person or stakeholder based on his or her role as a stakeholder and not as an expert per se. There is little or no direction to a solution provided by the facilitator, and the response is often qualitative in nature as a result, rather than the usual expert-driven quantitative output.

The strong emphasis on ownership addresses a key problem of the core techniques described earlier, and assists wider acceptance of outcomes at later stages. It is also viewed as complementary to more scientific methods with boundless scope for creativity. However, it has weaknesses in its emphasis on qualitative methods which can be difficult to record and difficult to justify at later dates, and it has been stated that the methods are still underdeveloped (ICIS 1999).

Judgmental forecasting therefore remains dominant despite the attempts at refinement listed above. Risk assessment, where the outcomes are subjected to a systematic testing to identify the potential likelihood, consequence and impact of possible problems and the sensitivity of solutions (CIRIA 1994), is a useful additional technique to add to all of the above methods at the output stage.

7.4 The information management angle

Having concentrated for much of the chapter on the individual's response to complex events and activities, it is useful to briefly make note of a school of thought which sees the complexity arising not so much from the events but from the information that arises.

It may be that, at the heart of the problem, lies information and the approach to information seeking, acquisition and use. Wilson and Walsh (1996) provide a good summary of information behaviour. This may provide better clues to explain why the current crop of working tools for 'harmonising' thought on the subject, while helpful, will not provide solutions to the problems or identify causes rather than effects. They suggest that information use has four main phases (Fig. 7.5): the intelligence phase when information is sought and gathered, the design phase when plans are developed to use the information, the choice phase when decisions are made and the review phase when the results are studied and feedback occurs.

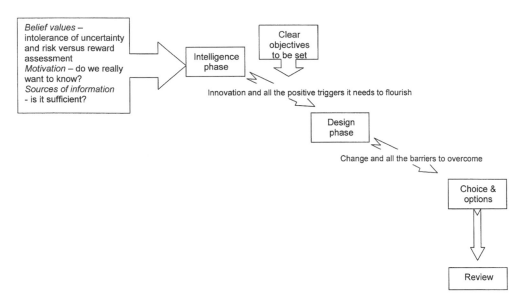

Fig. 7.5 The triggers in information use.

Each of these phases has a series of motivational factors and barriers associated with it. The motivational factors in the intelligence phase include the belief value of the information seeker, where past experience leading to gratification is a key factor, perceived risk, the desire for knowledge and outcomes and the intolerance of uncertainty also play a part. Barriers include the perceived risk associated with gathering the information and its weighting against any reward, and the perception of whether there is sufficient information available. Another problem is selective exposure, a tendency to select information that suits prior-held beliefs, attitudes and knowledge. The motivation and barriers at design phase hinge on innovation as a concept and all the triggers that this needs for success. At the choice phase the concept of change and all the barriers entailed in change management become apparent, causing further fear and avoidance.

Applying this to sustainable development one conclusion would certainly be that the intelligence phase is neither complete nor can it ever be fully complete. Figure 7.5 illustrates how Wilson and Walsh's theory draws similar conclusions to those developed in Section 5.4. Looking at the previous chapters there is evidence of all the factors identified above at the intelligence phase. Perhaps most important of these remains the uncertainty associated with the evidence base. Thus players choose different directions in assuming how design can be approached. With the information gathered to date, however, it is clear that many have already started designing solutions despite the lack of information, and designs will suffer unless the 'lost' elements are included in the equation.

Rifkind (2001), for example, has painted a fascinating picture of a global market economy which is moving towards a radically different system, dubbed the 'network economy', highly dependent on IT and business pursuing total product support. As noted in Chapter 4 this may help environmental concerns but is likely to raise many problems with social issues. Although Rifkind's picture presents one view of what might arise in the future, he raises the important issue that IT has brought new analytical and informational tools that will change approaches. However, there are still gaps and it is doubtful if these will ever be eliminated. Thus, risk will never be fully eliminated.

7.5 Summary

The conclusion from Chapters 5 and 6 is that sustainable development needs better cross-disciplinary working and an awareness of the psychological barriers associated with it, better change management particularly in complex situations and more use of good futurity tools to aid change management. Such approaches address uncertainty through consensus or a wider knowledge base, but they do not eliminate it.

In summary, the elements of learning and the barriers to learning have been studied in this chapter. The three factors of cross-disciplinary study, futurity and change aid understanding of the institutional and individual barriers to progressing sustainable development. There are a number of tools to help better understand and deal with all three. However, uncertainty in decision-making, a key variable which leads to unsustainable development, can never be eliminated. Risk lies at the root of the perception of many bad decisions, and fear of failure.

One route seeks to perfect the art of decision-making by perfecting the information available, although it is doubtful if perfection is an achievable goal. The other approach seeks to involve as many views as possible in the original decision so that ownership and responsibility for success are more widely spread, reducing the possibility of unforeseen constraints. Both of these, however, tend to deal with effects rather than causes.

8 Redefining the Debate

8.1 Summary

It is useful to review the progress of the arguments through the book, before trying to draw conclusions from the text. A developed-world view such as that provided is not a complete picture, but it does provide a starting point for further debate.

Chapter 1 provided an introduction to the debate, outlining the parameters and the variety of definitions of the terms. Many difficulties arise, with a key question remaining on whether sustainable development is a starting-point, process or end-goal. The conclusion was that better identification of the main components of the debate required a detailed look at the various schools of thought on the subject and the background to their stance in the debate.

Chapters 2–4 reviewed some of the theories in existence, grouping them into economics-led views, environmentally led views and a sample of others, many of which lie between the other two. It is important that a rounded picture is studied, although many readers will already hold a view within this range. The evidence in support of the various schools was studied. The conclusion to this was that no one group holds all the rights to correct information, and much of the evidence is blurred. A key outcome was the issue of 'evidence threshold', a problem that applies to all schools, and it was noted that this is an area that needs more study. There is no immediately obvious link-up between many of the schools of thought, raising the question of whether sustainable development is a workable, consistent concept.

In Chapter 5 attention turned to practice and the practical interpretations of the debate. There is a strong belief in many circles that theory will follow practice rather than practice follow theory (Fig. 8.1). Evidence emerging from fieldwork and case studies revealed difficulties with scale and repeatability in translating practice-led work into a universal theory. It was concluded that practice-based theory creates as many caveats, constraints and barriers as it provides answers. Complexity abounds, which is difficult to both represent and simplify.

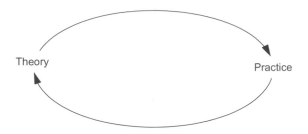

Fig. 8.1   The theory-practice circle.

The combination of evidence from Chapters 2–5 should have covered most of the issues viewed as important in the sustainable development debate. However, it is clear that some issues are ignored by some of the schools, some are accidentally missed and other issues may be too difficult to address within the context of each school's main premises for discussion.

Chapter 6 therefore explored this further, concentrating on the gaps in the debates. Seven important factors were highlighted as 'lost' in much of the current debate. These were viewed as significant omissions, which cause the current debate to be unbalanced. A second conclusion was that sustainable development also appears to be about complexity (of situation, of information and of group decision-making). This has a significant impact on human behaviour and perceptions, both of which are then studied in Chapter 7.

Chapter 7 majors on the key factors of cross-disciplinary approaches, change as a concept and futurity. It emerges from this work that cross-disciplinary work raises issues of complexity and change needs clear objectives. Futurity tools are available to help with these issues, but they cannot eliminate the risk and provide the certainty necessary to consistently produce good sustainable development decisions.

Thus, the conclusion so far is one of good intention from all sides, but all within a context of bias and neglect of the evidence that does not sit comfortably within one's own sphere of interest. This chapter takes this thread of thought forward and looks at why the concept still causes problems, why it must progress in the end and what it would take to be done before a clear way forward was possible.

8.2 Why does the concept cause problems?

On the whole, the idea of sustainable development in its various forms should strike a chord with everyone. Development, where it is necessary, that is sustainable is to be preferred to development that is unsustainable. The various definitions of sustainable development which were discussed in Chapters 2–5 cause confusion, but they each separately contain elements of an advertising executive's dream:

Mom and apple pie
Free lunch for the poor
Looking after the great outdoors
Everyone's a winner
Sons', and daughters' futures, etc.

Yet the concepts of sustainable development and sustainability do not and, in fact, may not ever receive universal acceptance. The US government's dismissal of the Kyoto agreement on climate change, a small part of sustainable development, concluded that it was unworkable and business would suffer (Sullivan 2001). Why? Part of the explanation lies in Chapters 6 and 7 which indicate that some of the more difficult issues are ignored by each of the schools in favour of easy answers. There are also significant complexities of learning associated with opening up thinking and approaches beyond the traditional single-discipline type approach. At this point in time, it can be argued that the complexity does complicate matters, making decision-making more difficult, with no obvious quick improvement as a result. Definitions, indicators and evidence bases all represent further serious barriers to progress and consensus, as does the legacy of past failures. The starting points of individuals, groups and even nations vary enormously in terms of inherited agendas which skew their ability to deal with new issues and inevitable change.

Sustainable development in the early debates was a largely environmental issue, and much of the public still sees it as such. The early targets for derision were industrialists who became scapegoats for all of the world's evils. While definitions have moved on, the industrial sectors have often switched off, and were happy to feed the US administration with a sceptical view on Kyoto, part truth part evidence-based.

Compromise will continue to be a major sticking point. For all to move over to a win–win situation all must accept a little loss, as was noted in the SUSPLAN project (2001) (Fig. 8.2).

The loss in the first step

Part of the SUSPLAN project (see Chapter 5) studied the perceptions of players in the planning process. Across a range of case studies it was clear that all players believe that the first move towards an agreed path to sustainable development results in a loss for themselves. While the long-term gain may be positive for all, most players have difficulty seeing beyond their own initial first step in the process, which is often a painful loss.

In Denmark, for example, the research concentrated on pig farmers who were having to clean up their operations because the groundwater of the area was becoming overpolluted. The saving for the community as a whole far outweighed the individual's loss. However, the saving was not immediate and in the short term the main result was the farmer's visible loss. In Newcastle a green field development involved compromise from all sides. However, few of the parties involved see the other party's loss and assumed that it was only themselves who had suffered.

Thus advocates of sustainable development often delude themselves by concentrating on the overall gain and ignoring the initial loss. In the grand scheme of things this initial loss may be relatively unimportant, but it is immediate and often highlighted by the individual players concerned. Ignoring or dismissing this anguish gives rise to resistance, often because the time path of projects leaves the initial loss with one party or group of people. At the time of this loss they see only themselves suffering the consequences.

Much development, as it operates today, involves a competition where the desired end point is to win at minimum cost (the starting point in Fig. 8.2). The cost is measured within the organisation. If that outcome involves others winning at minimum cost then that is an added bonus to society but of little consequence to the organisation. If others lose and they incur costs then it is a purely unfortunate consequence which is not factored into the equation.

In the early 1990s the concept of quality management was heavily marketed as the way forward for business just as sustainable development is now promoted as the way forward for society. It was based on the idea of 'get it right first time' or not

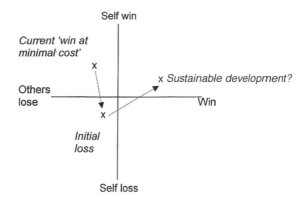

Fig. 8.2 The move to win–win from individual win at minimum cost.

passing the cost through the supply chain (Holmes and Willstead 1992). In a sense this caused the organisation to examine something more akin to its true costs in its near environment, i.e., within the groups close to the organisation. Thus, for example, the needs and costs of suppliers and clients were brought into a fuller cost equation for production companies.

Quality management led to assessment of mistakes and how systems in business accounted for them in a manner that put off the problem rather than addressing it. This resulted in a cost further down the line, often a cost split with the client. It was felt that if mistakes could be avoided within the system then the cost of failure could be avoided, providing a saving for the client but, ultimately, for the organisation itself, both directly through avoiding rectification and indirectly through not losing unhappy customers. Interestingly, this is a concept that involves an added initial cost at design and production stages (Fig. 8.3) and yet still was accepted throughout industry. The total cost is the sum of the

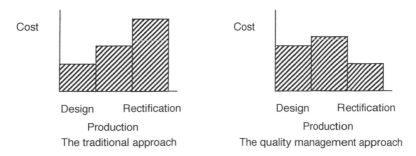

Fig. 8.3 The differing cost patterns associated with quality management.

three blocks, which is assumed to be least for the getting-it-right-first-time approach. This has a benefit for the company itself, the suppliers (if they are involved) and the client who receives a better product or service. Hawken *et al.* (1999) approach the same subject from a slightly different angle using the Japanese concept of 'muda' or waste and the rigorous removal of it from the supply chain as their core argument.

Sustainable development is a logical progression of all of this, extending the boundary further into the far environment (Fig. 8.4). It calls for organisations to examine whether they are passing on costs to anybody else or any place else, rather than just somebody they know. This extension from somebody or some place we know to any place is a difficult concept. The solution is easier stated than actioned and the issue of where to draw the boundary causes clear problems.

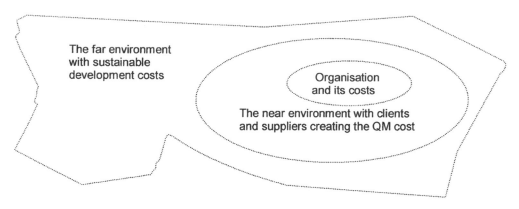

Fig. 8.4   The layers of operational environment for organisations.

Two solutions to this difficult question are often put forward, as noted in previous chapters. From an efficiency (and possibly the economist) viewpoint the solution lies in the whole-life costing direction and 'getting it perfect' first time. The focus is the technical correctness of the product. For the social scientist the solution advocated involves consulting others before getting it right first time. The focus is ownership and bringing stakeholders into the decision.

Neither route is perfect since consultation may come up with something that is inefficient but comfortable while whole-life costing may lead to efficient but unpopular products. Both represent proxies of sustainable development rather than a true representation since neither eliminates risk.

Discussion at the SUSPLAN conference (2001) on modelling sustainable development came to similar conclusions. It pointed to a planners' dilemma in whether they use tools (with all their limits, fallibilities and bias) to present people with choice or should the starting point be choices and then develop tools to suit. The solution, which adds expense, would be to do both iteratively (Fig. 8.5).

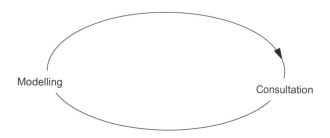

Fig. 8.5   The modelling–consultation circle.

Accepting the 'anybody, any place' argument would involve businesses thinking in terms of market size and not just market share, and having to take into account, for example, the natural resource implication of their market. It is interesting in this respect to note the methods of oil companies, frequently the targets of ridicule in sustainable development circles. They have often measured their wealth in terms of known natural reserves (Fig. 8.6) and have been ahead of the game in many respects. As a result, they are moving into renewable energy, managing change and

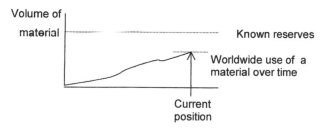

Fig. 8.6   Gap analysis of accumulated use of a material versus its natural limit.

developing scenario planning, all the result of keeping an eye for the future.

There is evidence that other businesses are also taking this message on-board. The *Economist* (2000) charts the change in attitude of business to global warming. From derision and disbelief that greenhouse gas effects existed in 1997 the *Economist* suggests the emphasis moved to complaint at the Hague Conference on Climate Change that governments were unable to agree an action plan on how to deal with the problem. Many big businesses, typically in sectoral groups rather than individually, now want clear ground-rules, development schemes and new market initiatives to improve the situation and foster innovation.

Working out the cost of a product or development to humanity or to a wider field of stakeholders makes for a complex analysis, and the cost of the analysis may be more than can be borne by the wider service, product, process or wherever it leads. So there is a need to have sensible debate on the subject. How do we draw the boundaries? How do we draw new plans for communicating along the project plan so that costs are shown to be shared and benefits are shown to be shared? Clearly this is a challenging agenda.

8.3   A response to the four main questions

It is important at this point to reconsider the four important questions raised at the start of Chapter 1. To date they have not been answered. Table 8.1 provides three sets of responses to the questions, representing some interesting insights from interested players and the views of the author.

The first responses arise from interpretation of discussion at the SUSPLAN conference, held in Newcastle, UK, in August 2001. This attracted a mixed audience of practitioners, politicians and academics looking at sustainable development in planning. The views reflect the mix and highlight the differences in language and culture of the broad groups of participants.

An interesting interpretation of question one appeared to develop at the SUSPLAN conference. In discussion it emerged that the politicians saw it as an end-goal or vision, the academics appear to see it as a starting-point, while the practitioners were caught in the middle trying to come to some sort of arrangement which might become a process.

Table 8.1 The four key questions.

The question	View from SUSPLAN (2001)	View from Sustainable Information Society (2001)	Author's view
Is it the starting-point, process or end-goal?	Politicians see it as end-goal, academics see it as starting-point, practitioners see it as process	Viewed as an end-goal (a workshop outcome)	Prefer to see it as process but more likely to be starting-point at present
How do we balance our developments?	A mix of consultation and modelling (see Fig. 8.5)	Leave it to the experts (as explained in Section 7.3)	Should be more systematic but must have scope for political decision
Is it a workable concept?	Yes (all agreed)	Yes (all agreed)	Yes, but needs refinement with three possible avenues (Table 6.3)
Can the best bits of theory be linked coherently?	Not addressed	Economic and environmental now linked but social needs special attention (Rifkind 2001)	Not at present (but see Fig. 8.9)

The SIS conference, held in Kouvola, Finland, in September 2001, attracted another mixed audience although mainly of academics and staff from large global organisations such as the United Nations, World Bank and European Commission. The key theme was the information society and its part in sustainable development. The views from this conference reflect a different, often 'global expert' type of response.

The author's views on the four questions are based on the development of the discussion throughout the text, and are further explained below.

8.3.1 Starting-point, process or end-goal?

In Chapter 1 the question is raised, and all three options are considered and briefly reviewed. The difficulties of finding common definitions at both the start and end-goal points, establishing repeatability at the process stage and the fact–political decision mix are such that it appears that there is no simple category in which sustainable development might easily fit.

In Chapter 5 the question is addressed again and the subjects of process and end-goal are studied in more depth. The chief difficulties appear to be in the creation of a process that encompasses displacement and scale and the currency of measurement. These factors are obstacles in developing a simple answer to the key question.

In Chapter 6 the 'process' angle was ignored in favour of looking at how principles, the starting-point, are linked to indicators, the frequent proxy for end-goals. This appeared to be a useful short-cut, but it has its limitations, with consistency of approach the primary problem.

The principles are reasonably well established, although some inconsistency remains, supporting the viewpoint that it fits best into the category of a starting-point at present, a vision for the future. However, the discussion earlier in Chapter 8 about the parallels with quality management systems suggests that a process type of model explanation may be possible in the future, although it is likely to produce a very complex model.

8.3.2   How do we balance developments?

Discussion in Chapter 1 and again in Chapter 5 suggests that, while balance across social, economic and environmental values is a laudable aim, it does not have a practical achievable concrete equivalent which lends itself to a SMART target. The better alternative, first suggested in Fig. 1.8, may be to separate out the hard evidence from the political decision in a staged decision, although this clearly needs further development. Three factors hold the key to making a more useful and coherent model. All three need simple, not complex tools for a workable approach:

(1) A systematic approach to decisions acknowledging the 'value judgement' versus the 'fact' components
(2) Proper identification and remedy to displacement problems
(3) Agreement on evidence thresholds for more complex subjects where damage thresholds and availability of evidence have different time or physical scales, to avoid damage.

8.3.3 Is sustainable development a workable concept?

This is the one area where there appears to be universal agreement. The views of practitioners, in accepting the concept, albeit on their own terms, are an influential factor. However, the lack of a transparent universal process of dealing with the subject still causes concerns. Any such universal process would need to be simple and practical despite the acknowledged complexity of the overall subject, a daunting challenge.

Reviewing the arguments presented throughout the text, there are three easily visualised avenues which offer a glimpse of hope for the future.

The scientific v. political v. unknown model

Figure 1.8 and further comment in Chapter 5 form the basis for one approach. This is a model that recognises the need for an evidence-base, recognises it is incomplete, accepts the need for a political decision and agrees that there are still unknowns (Fig. 8.7). It is probably the closest to the current system of judgement on development, although it is seldom as systematic or straightforward.

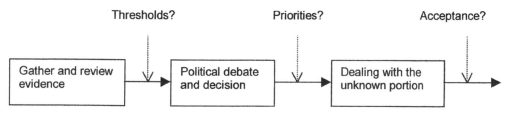

Fig. 8.7 The amended scientific v. political v. unknown model.

Clearly the issues that would need to be resolved to make it truly systematic are the evidence thresholds, political priorities and the means of accepting or dealing with the risk from the unknown portion. Such a model would fit with the steps suggested in Fig. 5.4 and follows the Wilson model on information management suggested in Fig. 7.5. As noted earlier in the text, however, while it may be easy to visualise or conceptualise such a process, it will be much harder to define the detail.

The quality management follow-on model

This line of argument originates in Chapter 4 and initially relies heavily on the arguments of Hawken *et al.* (1999). The traditional break-down of product development, service provision or other thinking into a set of single disciplines is queried and the concept of joining up the thinking, product development and service provision results. This indirectly introduces quality management and whole-life costing principles (Fig. 4.2).

In Section 5.3 the need to think beyond the near environment is then examined, taking us beyond quality management and whole-life costing. The step beyond the near environment is not an easy one – it has been visualised in very simplistic manners with two triangles in Fig. 5.3 and, again, in a different form, in Fig. 8.4. It will represent a major challenge to develop a coherent, logical model such as that illustrated conceptually in Fig. 8.8. Furthermore, as with quality management and whole-life costing before it, it will entail initial losses for many key players, which will cause initial reticence to get involved, as was illustrated in Fig. 8.2.

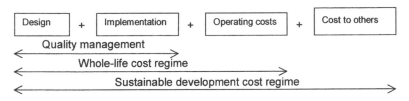

Fig. 8.8 The quality management follow-on model.

It will be very difficult to cover all effects in the far-field and therefore decisions will have to be made about sensible boundaries. However, viewed in this manner, there is scope for progress.

The principle to indicator model

This appears to be the preferred route at present for many players. However, as indicated in Chapter 1 and again in Chapter 6 it is often very difficult to see the relationship between choice of indicator, principles and best practice in indicator theory.

The analysis in Chapter 6 puts forward three options based on three sets of principles (Table 6.3); these rely on stakeholder involvement, balanced coverage and systematic review respectively. These represent the most obvious short-term way forward despite the very obvious limitations of each approach. Work remains to be done on this type of approach, and a hybrid method will probably emerge from on-going work.

The three approaches outlined above represent the starting-point for attempts to link up the important components of sustainable development. They may help to provide different options depending on the method chosen, which will form the basis for decisions on the type of balance desired from future development. However, even in a more refined format, they will never eliminate risk and neither do they address the key problem of predicting the future.

8.3.4   Can a coherent theory be developed?

A view on whether sustainable development will ever form a coherent theory hinges on linking all three elements (i.e., principles, elements and indicators) coherently (see Fig. 6.1). The views from within traditional economics circles, that a little tinkering around the edges of their own subject will suffice, are not persuasive. Concerns about the narrow evidence-base and the basic assumptions involved are widespread. The reformers within this school (Hawken *et al.*, Pearce, Rifkind etc.) have helped to identify the problems with futurity and the misalignment with environmental and social issues.

New environmental markets (Chapter 2), thoughts that social capital is special and needs nurturing (Chapter 4) or that new forms of leasing rather than ownership (Chapter 4) will mean that environmental concerns will be more closely woven with the economic system do not seem fundamental enough to address any step change necessary. These issues, however, need further research and a higher priority than they are accorded within the two-dimensional thinking of current economic models. The environmental schools, by contrast, have shown that the environment has problems that need addressing but have yet to show a coherent theory to link the environment with the social and economic. They likewise have yet to make the leap fully into the complex issue of creating a single coherent theory, which addresses current and future economic, social and environmental concerns.

The view of Henk Voogd, noted in Chapter 1, provides an

important contribution to the debate. The big challenge is to stretch the players to take the wider view and embrace a discussion that incorporates all of the key variables, including those outlined in Chapter 6.

While complexity will undoubtedly be an issue for all of the players, Fig. 8.9 represents an interpretation of the debate to date which offers some hope for the future. Using the time-scales suggested, it may be that more experimentation and discussion are required before a coherent theory is developed, similar to the conclusions that arose from discussion around Fig. 7.5 on the triggers in information use.

This would lead one to believe that sustainable development remains a contested theory for the present, but with the prospect of progress in the future. A set of principles or assumptions and a set of evidence are the starting-point of any coherent theory, and the earlier chapters indicate that work has started on these.

8.4   The way forward: placing the frameworks in their context

The broadest implication of the sustainable development debate is the belief that the world had reached a point where it was necessary to discuss the way forward. Change is a difficult process, made more difficult because (1) the pace of change has increased, making it difficult to make good decisions since there is little time for analysis and digestion, even though more information is available for decision-making, and (2) changes in information management have forced a broad movement to more participative forms of planning.

In parallel with this desire to talk through change, the development of the science and monitoring associated with the environment has shown alarming situations which may be a foretaste of bigger disasters to come. The most obvious example of this is climate change. As the debate on that particular issue has shown, the evidence and implications are unclear, allowing plenty of 'wriggle-room' for all participants.

It is often proclaimed that the goal of sustainable development is to perfectly align the three frameworks in a bull's eye configuration, as shown in the right part of Fig. 8.10. However, the left part of the figure deliberately shows the three frameworks in an off-centre manner, because perfect alignment has little or no meaning, and actually confuses our understanding of the balance required. The goal is to get the economic framework to fit somewhere inside the

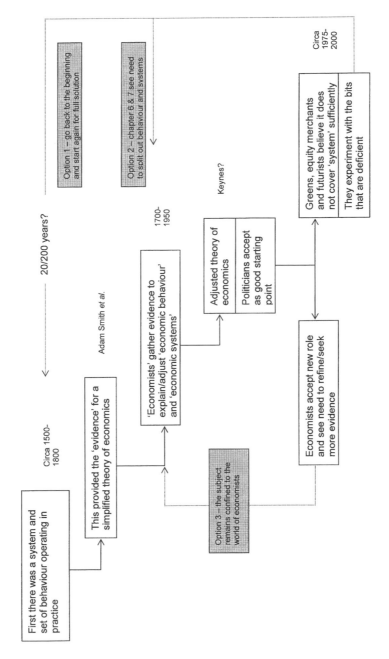

Fig. 8.9 The proper time-scales for Sustainable Development theories.

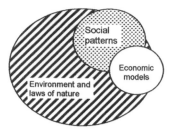
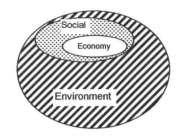

The current position   The best improvement?

Fig. 8.10   The misaligned bull's eye.

social framework, and the social framework to fit within the environmental, rather than centre them on some hypothetical point. A good comparison is the discussion in Chapter 1 on the triangle and the misleading belief that the centroid is the balance point. It is the starting-point and the process of reaching a satis-factory end-goal that are important in this particular model of the theory. What is clear is that they are misaligned at present. To complicate matters, the frameworks themselves are different; the economic framework is a series of rules representing a set of the-oretical models which assume participants' perfect knowledge, the social framework is based on patterns of behaviour which are described from observation and the environmental framework has, at its core, the 'laws' of nature but with many pockets of uncertainty.

The currency for interpretation across the three frameworks therefore presents many difficulties. Economic choices do not match social and environmental issues and vice versa, and to base the currency on the most limited framework, the economic model, is clearly problematic. Other 'proxies' can and do work, but they are an imperfect tool.

8.5   Conclusions

The link between starting-point, the process and the end-goal has been an unsolvable simplistic query throughout the text. A wider debate with inclusion of the seven factors discussed in Chapter 7 would represent a significant improvement to the overall sustainable development discussion, but there are still many other gaps to address.

It is encouraging to note that where there is debate then there can be progress. It has already been noted that the tools, which can be employed to analyse, provide data and inform the public, are still to be fully developed. For example, in the SUSPLAN project the diverging needs of the three partner countries led to different improvements with a common goal. The Danes showed how Geographical Information Systems are a useful tool to start the debates, while the Dutch and British showed how mapping perception, which differs so significantly from player to player, helps to progress the debate. A model of communication is needed that allows the players to see the benefits for all and to map out when their own individual benefit will arrive.

However, the major issue remains the lack of clear evidence for decisions on development, which was first raised in Table 3.5, and the risk that arises as a result of this. The lack of clear evidence often leaves humanity with a choice – act before the event using preventative principles or leave it until the evidence is available. Just as in law, which has similar debates, the sustainable debate contains advocates of both systems. It is often the first people involved in the discussion (or those with the strongest position) who dictate the terms of the follow-up debate. The implications of this are enormous for nations, businesses, communities and individuals. What is clear is that change is necessary. What is not clear is how much change is necessary, leaving the way open for the debate on how much is needed or desirable.

Current evidence suggests that Europe is leaning towards preventative approaches while the USA appears to lean towards more evidence-first approaches. If that is the case then the USA will escape the initial costs that were unnecessary and may learn from the early lessons of Europe. However, it may have a bigger clean-up bill, in the manner shown from quality management arguments suggested in Sections 8.2 and 8.3. In Europe the debate may have skewed away from business as the greens have captured the political and moral high ground and will impose preventative measures against individuals and businesses. In the USA the Bush administration appears to tend towards letting business and government bypass even the serious issues that must be addressed.

It may be that the much-abused concept of compromise will feature highly in the ultimate solutions. Sustainable development will have its day as a serious subject because change is ultimately necessary and it is a subject that humanity increasingly needs to get right.

The debates at national level and, to a certain extent, individual life-style level are interesting but not quite as fascinating as at the community level – the level at which all the competing interests meet. The challenge of mixing and matching and getting it right at this level is a huge reason for the advocates of sustainable development to continue their search for better forms of progress, and for practitioners to continue to play a prominent part in the progress of the debate and the evolving subject.

References

References have been split by chapter to ease sourcing material for the reader. This has entailed some references being repeated in a number of chapters.

Chapter 1

Brundtland, G. (1987) World Commission on Environment and Development. *Our Common Future*. Oxford University Press, Oxford.

Cantle, T. (1999) Breaking the deadlock. *Town & Country Planning*, **68**(8), 254–5.

Coates, J.B. *et al.* (1993) *Corporate Performance Evaluation in Multinationals*. Chartered Institute of Management Accountants, London.

Department of Environment, Transport and Regions (1999a) *A Better Quality of Life: A Strategy for Sustainable Development for the United Kingdom*. DETR, London.

Department of Environment, Transport and Regions (1999b) *Quality of Life Counts*. DETR, London.

Economist (1999) The end of urban man? Care to bet? 31/12/99, pp. 31–2.

Economist (2001) No time to plod. 27/10/01, pp. 64–6.

European Commission (1996) *European Sustainable Cities*. Expert Group on the Urban Environment, Luxembourg.

Girardet, H. (1999) *Creating Sustainable Cities*. Greenspan, Totnes, Devon.

HM Treasury (1997) *Appraisal and Evaluation in Central Government*. HMSO, London.

International Council for Local Environmental Initiatives (1994) *The Local Agenda 21 Initiative – ICLEI Guidelines for Local and National Local Agenda 21 Campaigns*. ICLEI, Toronto.

IUCN *et al.* (1991) *Caring for the Earth: A Strategy for Sustainable Living*. IUCN, Gland, Switzerland.

Krupp, H. (1996) Sustainable development: the unsolved dilemma, is there a way out? *Environmental Technology for Northern Europe*, Hagbarth, Sweden.

Local Government Association (1999) *Local Sustainability Counts: A*

Guide to Core Set of Local Quality of Life Indicators, Draft 1: CLIP Indicators Handbook. LGA, London.

Local Government Management Board, UK (1993) *A Framework for Local Sustainability*. LGMB, Luton.

Mawhinney, M. (1999) Private sector partnerships workshop presentation at *Review of Function of Regional Development Agencies*. Planning Exchange, Manchester.

Murray, S. (2001) The message gets muddled. Responsible business in the global economy, *Financial Times*, October 2001, London.

National Stategies for Sustainable Development (2000) (on-line) http://www.nssd.net/index1.html.

Novartis Foundation for Sustainable Development (2001) *Welcome Page* (on-line) http://www.foundation.novartis.com.

Pearce, D. *et al.* (1990) *Blueprint for a Green Economy*. Earthscan Publications, London.

Pezzey, J. (1989) *Economic Analysis of Sustainable Growth and Sustainable Development*. World Bank Environment Department Working Paper no. 15, May 1989, Washington.

Robert, K.H. *et al.* (1997) A compass for sustainable development. *International Journal of Sustainable Development and World Ecology*, 4(2), 79–92.

Schoonbrodt, R. (1995) *The Sustainable City: Part 2, the SMEs and the Revitalisation of European Cities*. European Foundation for the Improvement of Living and Working Conditions, Dublin.

US Department of Energy (2001) *Center of Excellence for Sustainable Development* (on-line) http://www.sustainable.doe.gov/overview/ovintro.shtml.

Voogd, H. (2001) Keynote presentation at conference opening *SUSPLAN 2001. Conference Proceedings*, Newcastle.

Wackernagel, M. & Rees, W.E. (1996) *Our Ecological Footprint*. New Society Publishers, Gabriola Island, British Columbia.

World Bank (2001) *Sustainable Development* (on-line) http://www.worldbank.org/depweb/english/whatis.htm.

World Business Council for Sustainable Development (2001) *What Is the WBCSD?* (on-line) http://www.wbcsd.ch/aboutus/index.htm.

Chapter 2

Appleyard, B. (2001) Mind games. *Sunday Times*, News Review section, 29/4/01, p. 4.

Berry, L.M. & Houston, J.P. (1993) *Psychology at Work*. WCB Brown & Benchmark, Los Angeles.

Daly, H.E. & Cobb, J.B. (1989) *For the Common Good: Redirecting the Economy Towards Community, the Environment and a Sustainable Future*. Green Print, London.

Dunne, N. (2001) White House split as Bush rejects Kyoto. *Financial Times*, 30/3/01, p. 6.

Economist (1999) The non-governmental order. 11/12/99, pp. 22–4.

Economist (2000a) The World Bank's muddled prescriptions. 30/9/00, p. 130.

Economist (2000b) Anti-liberalism old and new. 21/10/00, p. 146.

Economist (2000c) Anti-capitalist protests. 23/9/00, pp. 125–9.

Economist (2000d) The ethics gap. 2/12/00, p. 132.

Economist (2000e) The real losers. 11/11/00, p. 15.

Economist (2000f) Hotting up in the Hague. 18/11/00, pp. 133–6.

Economist (2000g) The health effect. 3/6/00, p. 124.

Economist (2000h) Economics focus: growth is good. 27/5/00, p. 122.

Economist (2000i) Stumbling yet again? 16/9/00, pp. 117–22.

Economist (2000j) Beyond the Hague. 2/12/00, pp. 23–4.

Economist (2000k) Self-centred. 8/7/00, pp. 134–5.

Economist (2000l) Running the numbers. 14/10/00, p. 152.

Economist (2000m) Productivity on stilts. 10/6/00, p. 130.

Economist (2000n) The bigger they are. 28/10/00, pp. 115–21.

Economist (2000o) Morality pays. 8/7/00, pp. 122–5.

Economist (2001a) Green and growing. 27/1/01, pp. 106–9.

Economist (2001b) The cutting edge. 24/2/01, p. 127.

Economist (2001c) Getting better all the time. Supplement, 10/11/01.

Economist (2001d) Oh no Kyoto. 7/4/01, pp. 95–8.

Economist (2001e) Managing the rainforests. 12/5/01, pp. 117–20.

Esty, D. *et al.* (2001) *Environmental Sustainability Index* (on-line) http://www.ciesin.org/indicators/ESI/esi.xls.

Hawken, P. *et al.* (1999) *Natural Capitalism*. Earthscan Publications, London.

Hopwood, B. (1999) *Notes on Calculation of GDP*. Internal working paper produced at Sustainable Cities Research Institute, University of Northumbria, Newcastle.

Jackson, T. *et al.* (1997) *Sustainable Economic Welfare in the UK*, 1950–96. New Economics Foundation, London.

Martin-Fagg, R. (1996) *Making Sense of the Economy*. International Thomson Business Press, London.

Meadows, D. *et al.* (1992) *Beyond the Limits: Global Collapse or a Sustainable Future*. Earthscan Publications, London.

Middleton, N. *et al.* (1993) *Tears of the Crocodile*. Pluto Press, London.

Mills, R. *et al.* (1995) (*Managing Resources: 4. Managing Economics.* HDL Training and Development Limited, Henley.

Pearce, D. *et al.* (1990) *Blueprint for a Green Economy.* Earthscan Publications, London.

Pfefferman, G. (2001) Poverty reduction in developing countries. *Finance and Development*, June 2001, pp. 42–5.

Schott, J. (2000) After Seattle. *Economist*, 26/8/00, p. 90.

Skidelsky, R. (2000) Ideas and the world. *Economist*, 25/11/00, pp. 131–4.

Streeten, P. (2001) Integration, interdependence and globalization. *Finance and Development*, June 2001.

Thomas, V. (2000) Why quality matters. *Economist*, 7/10/00, p. 142.

Chapter 3

Appleyard, B. (2001) Mind games. *Sunday Times*, News Review section, 29/4/01, p. 4.

Brewis, K. (2001) If you are a child living by the Caspian Sea, venturing outside is a dangerous game to play. *Sunday Times*, Magazine section, pp. 32–8.

Budiansky, S. (1995) *Nature's Keepers: The New Science of Nature Management.* Weidenfeld and Nicholson, London.

Brundtland, G. (1987) World Commission on Environment and Development. *Our Common Future.* Oxford University Press, Oxford.

Daly, H.E. & Cobb, J.B. (1989) *For the Common Good: Redirecting the Economy Towards Community, the Environment and a Sustainable Future.* Green Print, London.

Dobson, A. (1995) *Green Political Thought.* Routledge, London.

Dunne, N. (2001) White House split as Bush rejects Kyoto. *Financial Times*, 30/3/01, p. 6.

Economist (2000a) The ethics gap. 2/12/00, p. 132.

Economist (2000b) Anti-liberalism old and new. 21/10/00, p. 146.

Economist (2000c) Anti-capitalist protests. 23/9/00, pp. 125–9.

Economist (2001) Climate change – getting real. 17/3/01, pp. 120–21.

Girardet, H. (1999) *Creating Sustainable Cities.* Greenspan, Totnes, Devon.

Goldsmith, E. (1992) *The Way.* Rider, London.

Goudie, A. (2000) *The Human Impact on the Natural Environment.* Blackwell, Oxford.

Hawken, P. *et al.* (1999) *Natural Capitalism.* Earthscan Publications, London.

The Holy Bible (1995) Cambridge University Press, Cambridge.

Hopwood, B. (2001) *Notes on Environmental Effects of Pollution*. Internal working paper produced at Sustainable Cities Research Institute, University of Northumbria, Newcastle.

Hopwood, B. *et al.* (2000) *Sustainable Development: Mapping Different Approaches*. Paper presented at the Second Sustainable Cities Network Conference, Manchester, September 2000.

International Water Management Institute (2000) *Projected Water Scarcity in 2025* (on-line) http://www.cgiar.org/wmi/home/wsmap.html.

IUCN *et al.* (1980) *World Conservation Strategy: Living Resource Conservation for Sustainable Development*, Gland, Switzerland.

IUCN *et al.* (1991) *Caring for the Earth: A Strategy for Sustainable Living*. IUCN, Gland, Switzerland.

Jacob, R. (2001) Asia propelled to brink of environmental catastrophe. *Financial Times*, 30/3/01, p. 19.

Leopold, A. (1966) *A Sand County Almanac: With Essays on Conservation from Round River*. Oxford University Press, New York.

Lomberg, B. (2001) The truth about the environment. *Economist*, 4/8/01, pp. 71–3.

Lovelock, J. (1988) *The Ages of Gaia: A Biography of Our Living Earth*. Oxford University Press, Oxford.

Meadows, D. *et al.* (1972) *The Limits to Growth: A Report for the Club of Rome's Project on the Predicament of Mankind*. Earth Island, London.

Mellor, M. (1997) *Feminism and Ecology*. Polity, Cambridge.

Middleton, N. *et al.* (1993) *Tears of the Crocodile*. Pluto Press, London.

Murray, S. (2001) The message gets muddled. Responsible business in the global economy. *Financial Times*, October 2001.

Mylius, A. (2000) Keeping the peace. *New Civil Engineer*, 20/4/00, pp. 14–15.

Naess, A. (1989) *Ecology, Community and Lifestyle*. Cambridge University Press, Cambridge.

O'Riordan, T. (1989) The challenge for environmentalism. In: *New Models in Geography* (eds Peet, R. & Thrift, N.). Unwin Hyman, London.

Pearce, D. *et al.* (1990) *Blueprint for a Green Economy*. Earthscan Publications, London.

Schumacher, E. (1973) *Small Is Beautiful: Economics as if People Mattered*. Abacus, London.

Smith, D. (1999) *The State of the World Atlas*. Penguin, London.

Spence, M. (1999) *Dispossessing the Wilderness: Indian Removal and the Making of National Parks*. Oxford University Press, New York.

Sullivan, A. (2001) Lizzie crashes into America's class war. *Sunday Times*, News Review section, 29/7/01, p. 6.

United Nations Environmental Programme (1999) *Global Environmental Outlook*. Earthscan Publications, London.

Wackernagel, M. & Rees, W.E. (1996) *Our Ecological Footprint*. New Society Publishers, Gabriola Island, British Columbia.

Weizacker, E.V. *et al.* (1998) *Factor Four*. Earthscan Publications, London.

Wood, D. (1995) Conserved to death. *Land Use Policy*, **12**(2), pp. 115–35.

World Bank (2001) *World Development Indicators* (on-line) http://www.worldbank.org/data/wdi2001/index.html.

World Wildlife Fund (2000) *Annual Review*. WWF-UK, Goldalming.

Chapter 4

Commission of the European Communities (2001) *Consultation Paper for the Preparation of the European Union Strategy for Sustainable Development*. CEC, Brussels.

Driscoll, M. (2001) No peace for the pirates. *Sunday Times*, News Review section, 1/4/01, p. 4.

Economist (1999) The non-governmental order. 11/12/99, pp. 22–4.

Economist (2001a) The strange persistence of politics. 31/3/01, p. 41.

Economist (2001b) What next then? 28/7/01, p. 77.

Economist (2001c) Oh no Kyoto. 7/4/01, pp. 95–8.

Hawken, P. *et al.* (1999) *Natural Capitalism*. Earthscan Publications, London.

Hopwood, B. *et al.* (2000) *Sustainable Development: Mapping Different Approaches*. Paper presented at the Second Sustainable Cities Network Conference, Manchester, September 2000.

Middleton, N. *et al.* (1993) *Tears of the Crocodile*. Pluto Press, London.

Mies, M. & Shiva, V. (1993) *Ecofeminism*. Zed Press, London.

One World Action (2001) *About the Political Compass* (on-line) http://64.224.73.234/politicalcompass/analysis2.html.

Pearce, D. *et al.* (1990) *Blueprint for a Green Economy*. Earthscan Publications, London.

Pepper, D. (1993) *Ecosocialism: From Deep Ecology to Social Justice*. Routledge, London.

Rifkind, J. (2001) Sustainable Information Society. Paper presented to the *Sustainable Information Society – Values and Everyday Life* Conference, Kouvola, Finland.

Sullivan, A. (2001) It's sweet and sour for America's poor. *Sunday Times*, News Review section, 3/6/01, p. 8.

Wade, R. (2001) Winners and losers. *Economist* 28/4/01, pp. 93–7.

World Bank (2001) *World Development Indicators* (on-line)
http://www.worldbank.org/data/wdi2001/index.html.

Chapter 5

Carnall, C. (1995) *Managing Change in Organisations*, 2nd edn.
Prentice-Hall, Hemel Hempstead.

Denton, G. (1981) Regional divergence in the community with
special reference to EU enlargement. In: *Economic Divergence in
the EC* (eds M. Hodges & W. Wallace). Allen and Unwin,
London.

Department of Environment, Transport and Regions (1999) *A
Better Quality of Life: A Strategy for Sustainable Development for the
United Kingdom*. DETR, London.

DG-Environment (2000) *Towards a Local Sustainability Profile: European Common Indicators*. European Commission, Brussels.

European Commission (1996) *European Sustainable Cities*. Expert
Group on the Urban Environment, Luxembourg.

Girardet, H. (2000) Cities People Planet. A presentation to the
3rd Euoprean Conference on Sustainable Cities and Towns,
Hannover.

Hawken, P. *et al.* (1999) *Natural Capitalism*. Earthscan Publications,
London.

HM Treasury (1995) *Framework for the Evaluation of Regeneration
Projects and Programmes*. HMSO, London.

Jackson, T. *et al.* (1997) *Sustainable Economic Welfare in the UK,
1950–96*. New Economics Foundation, London.

Mawhinney, M. (2000) Pitfalls and constraints in the study of
sustainable cities. *Cityscape–Landscape Conference Proceedings*,
Carlisle.

Pearce, D. *et al.* (1990) *Blueprint for a Green Economy*. Earthscan
Publications, London.

Porter, G. (2000) *SUSPLAN: Developing Tools for Sustainable Planning: UK Activity Report for Phase 1*. Sustainable Cities Research
Institute, University of Northumbria, Newcastle.

Porter, G. *et al.* (2001) *A Common Approach to Developing Sustainability in the Spatial Planning Processes of Denmark, Netherlands and
the UK*. Sustainable Cities Research Institute, University of
Northumbria, Newcastle.

Richardson, K. (2001) Creating neighbourhoods in balance – issues
and solutions. *Man and City Conference*, Naples.

Rifkind, J. (2001) Sustainable Information Society. Paper presented
to the *Sustainable Information Society – Values and Everyday Life*
Conference, Kouvola, Finland.

Sustainability North East (2001) *Towards a Regional Framework*. Sustainability North East, Newcastle.

Chapter 6

Ahti, A. (2001) Welcome and opening presented to the *Sustainable Information Society – Values and Everyday Life* Conference, Kouvola, Finland.

Arup (2001) SpeAR: *A Sustainability Assessment Methodology*. Arup, London.

Best, A. *et al.* (1998) *Sustainable Seattle: Indicators of Sustainable Community*. A. Best, Seattle.

Chambers, N. *et al.* (2000) *Sharing Nature's Interest: Ecological Footprints*. Earthscan Publications, London.

Commission of the European Communities (2001) *Consultation Paper for the Preparation of the European Union Strategy for Sustainable Development*. CEC, Brussels.

de Boer, R. & de Roo, G. (2001) Coping with sustainability in a complex and intersubjective world: the evaluation of a planning model that focuses on desired, present and potential contribution of the actors involved. *SUSPLAN 2001 Conference Proceedings*, Newcastle.

Department of Environment, Transport and Regions (1999) *A Better Quality of Life: A Strategy for Sustainable Development for the United Kingdom*. DETR, London.

DG-Environment (2000) *Towards a Local Sustainability Profile: European Common Indicators*. European Commission, Brussels.

Esty, D. *et al.* (2001) *Environmental Sustainability Index* (on-line) http://www.ciesin.org/indicators/ESI/esi.xls.

Girardet, H. (1999) *Creating Sustainable Cities*. Greenspan, Totnes, Devon.

Girardet, H. (2000) *Cities People Planet*. A presentation to the 3rd European Conference on Sustainable Cities and Towns, Hannover.

Greenhalgh, P. *et al.* (2001) The response of industrial and office occupiers to property led regeneration policies in Tyne and Wear. *SUSPLAN 2001 Conference Proceedings*, Newcastle.

Hansen, B. (2001) Sustainable planning in practice. *SUSPLAN 2001 Conference Proceedings*, Newcastle.

Healey, P. (2001) Strategic planning for complex situations: new approaches to strategic spatial planning. *SUSPLAN 2001 Conference Proceedings*, Newcastle.

HM Treasury (1995) *Framework for the Evaluation of Regeneration Projects and Programmes*. HMSO, London.

Hopwood, B. *et al.* (2000) *Sustainable Development: Mapping Different Approaches.* Paper presented at the Second Sustainable Cities Network Conference, Manchester, September 2000.

Meadows, D.H. *et al.* (1972) *The Limits to Growth.* Universe Books, New York.

Mega, V. & Pedersen, J. (1998) *Urban Sustainability Indicators.* European Foundation for the Improvement of Living and Working Conditions, Dublin.

Middleton, N. *et al.* (1993) *Tears of the Crocodile.* Pluto Press, London.

Miller, G. (2000) *Living in the Environment*, 11th edn. Brooke/Cole Publishing, Pacific Grove.

Mumford, L. (1961) *The City in History.* MJF Books, New York.

OECD (2001) *Environmental Indicators: Towards Sustainable Development.* OECD, Paris.

Pearce, D. *et al.* (1990) *Blueprint for a Green Economy.* Earthscan Publications, London.

Robson, D. *et al.* (1998) *The Impact of UDCs in Leeds, Bristol and Central Manchester.* University of Manchester.

Streeten, P. (2001) Integration, interdependence and globalization. *Finance and Development*, June 2001.

Sullivan, A. (2001) Lizzie crashes into America's class war. *Sunday Times*, News Review section, 29/7/01, p. 6.

SUSPLAN (2001) Conference, Newcastle, August 2001.

Sustainable Information Society – Values and Everyday Life Conference, Kouvola, Finland, September 2001.

United Nations Development Programme (2001) *Human Development Index* (on-line)
http://www.undp.org/hdr2001/indicator.

World Bank (2001a) *Sustainable Development* (on-line)
http://www.worldbank.org/depweb/english/whatis.htm.

World Bank (2001b) *World Development Indicators* (on-line)
http://www.worldbank.org/data/wdi2001/index.html.

Chapter 7

Berry, L.M. & Houston, J.P. (1993) *Psychology at Work.* WCB Brown & Benchmark, Los Angeles.

Bullet Point (2001) *Why Change Fails. Sample Issue 2001.* Bulletpoint Communications, Redhill, p. 1.

Carnall, C. (1995) *Managing Change in Organisations*, 2nd edn. Prentice-Hall, Hemel Hempstead.

CIRIA (1994) *Control of Risk.* Special Publication no. 125. CIRIA, London.

Collins (1998) *Collins English Dictionary* (Consultant J. Sinclair). Harper Collins, Glasgow.

Eglin, R. (2001) Think ahead to avoid any nasty surprises. *Sunday Times*, Appointments Section, 29/4/01, p. 16.

Glenn, J. (2001) Millenium Project presented to the *Sustainable Information Society – Values and Everyday Life* Conference, Kouvola, Finland.

ICIS (1999) *Integrated Assessment: A Bird's Eye View*. ICIS, Maastrict University.

Kassler, P. (1995) Scenarios for world energy: barricades or new frontiers? *Long Range Planning*, **28**, December 1995, 38–47.

Mawhinney, M. (2000) Pitfalls and constraints in the study of sustainable cities. *Cityscape–Landscape Conference Proceedings*, Carlisle.

Mayank, H. (2000) Correspondence by e-mail on Bremen Partnership.

Moyer, K. (1996) Scenario planning at British Airways – a case study. *Long Range Planning*, **29**, April 1996, 172–81.

Palmer, M. (1998) *Business Transformation*. MBA Course Notes, Henley Management College.

Ravetz, J. (2000) *Integrated Visions NW-UK*. Centre for Urban and Rural Ecology, Manchester University.

Rifkind, J. (2001) Sustainable Information Society. Paper presented to the *Sustainable Information Society – Values and Everyday Life* Conference, Kouvola, Finland.

Royal Navy (1995) *Getting Things Done*. Royal Navy, Dartmouth.

Senge, P.M. (1996) *The Fifth Discipline: The Art and Practice of the Learning Organization*. Currency Doubleday, New York.

Sustainable Development Research Institute (2001) *George River Basin Project* (on-line) http://www.sdri.ubc.ca.

Tangram (2001) *The Manager's Toolkit Part 2: Looking After your Vitals* (on-line) http://www.tangram.co.uk.

van Asselt, M.B.A. *et al.* (1998) *Towards Visions for a Sustainable Europe*. ICIS, Maastricht University.

Wilson, T. & Walsh, C. (1996) *Information Behaviour: An Inter-Disciplinary Perspective* (on-line)
http://www.ukoln.ac.uk/services/papers/bl/blri010/.

Chapter 8

Economist (2000) Beyond the Hague 2/12/00, pp. 23–4.

Hawken, P. *et al.* (1999) *Natural Capitalism*. Earthscan Publications, London.

Holmes, A. & Willstead, C. (1992) *Internal Quality Auditing*. Lloyds Register Quality Assurance, London.

Rifkind, J. (2001) Sustainable Information Society. Paper presented to the *Sustainable Information Society – Values and Everyday Life* Conference, Kouvola, Finland, September 2001.

Sullivan, A. (2001) Why George W is right to reject Kyoto. *Sunday Times*, News Review section, 1/4/01, p. 6.

SUSPLAN (2001) Conference, Newcastle, August 2001.

Index

Note: tables and figures are in bold type; SD is *sustainable development*